Lecture Notes in Mathematics

Edited by A. Dold and B. Eckmann

1002

Albert Edrei
Edward B. Saff
Richard S. Varga

Zeros of Sections of Power Series

Springer-Verlag
Berlin Heidelberg New York Tokyo 1983

Authors

Albert Edrei
Department of Mathematics, Syracuse University
Syracuse, New York 13210, USA

Edward B. Saff
Center for Mathematical Services, University of South Florida
Tampa, Florida 33620, USA

Richard S. Varga
Institute for Computational Mathematics
Kent State University, Kent, Ohio 44242, USA

AMS Subject Classifications (1980): 30C15, 30D15, 30E15

ISBN 3-540-12318-0 Springer-Verlag Berlin Heidelberg New York Tokyo
ISBN 0-387-12318-0 Springer-Verlag New York Heidelberg Berlin Tokyo

Printing and binding: Beltz Offsetdruck, Hemsbach/Bergstr.
2146/3140-543210

The contribution of Albert Edrei is lovingly dedicated to *Lydia*.

The contribution of Edward Saff is lovingly dedicated to *Loretta* .

The contribution of Richard Varga is lovingly dedicated to *Esther*.

Acknowledgments

We sincerely wish to thank Mr. Amos J. Carpenter (now at the Department of Computer Science, St. Joseph's College, Rensselaer, IN 47978) who performed the necessary calculations for Figures 1-12. These calculations were done on a VAX-11/780 and graphed by a Calcom plotter (both at Kent State University).

We also express our gratitude to Faith Carver, Esther Clark, Anna Lucas and Mary McGill, who took great care in the typing of our manuscript.

Table of Contents

		Page
	Acknowledgments	
1.	Introduction	1
2.	Statements of our results	7
	I. Mittag-Leffler functions of order $\lambda > 1$	7
	II. Functions of genus zero whose zeros are all negative	19
	III. Problems for further study	23
3.	Discussion of our numerical results	26
	Figures 1 - 12	29-40
4.	Outline of the method	41
5.	Notational conventions	43
6.	Properties of the Mittag-Leffler function of order $1 < \lambda < \infty$	44
7.	Estimates for $G_m(w)$ and $Q_m(w)$	49
8.	A differential equation	53
9.	Estimates for $J_m(w)$ near the circumference $\|w\| = 1$ . .	60
10.	Existence and uniqueness of the Szegö curve	62
11.	Crude estimates for $\|U_m(w)\|$ and $\|Q_m(w)\|$	63
12.	Proof of Theorem 5	70
13.	Proof of Theorem 1	70
14.	Proof of Theorem 2	72
15.	The circular portion of the Szegö curve (Proof of Theorem 3)	77
16.	Proof of Theorem 4	80

Page

17. Proof of Theorem 6 82

18. Properties of $\mathcal{L}$-functions; proof of assertion I of Theorem 7 87

19. $\mathcal{L}$-functions of genus zero are admissible in the sense of Hayman 91

20. The functions $U_m(w)$, $Q_m(w)$, $G_m(w)$ associated with $\mathcal{L}$-functions of genus zero 92

21. Estimates for $U_m(w)$ 95

22. Determination of $\lim \Omega_m(\zeta)$ 99

23. Comparison with integrals; proof of assertion II of Theorem 7 101

24. The Szegö curves for $\mathcal{L}$-functions of genus zero 103

25. Estimates for $U_m(\sigma_m e^{i\phi} w)$ 106

26. Proof of assertion IV of Theorem 7 108

References 109

Index of Ad-Hoc Definitions and Notations 113

General Index 114

1. Introduction. Let $f(z)$ be an analytic function defined by its Taylor expansion

$$(1.1) \qquad f(z) = \sum_{j=0}^{\infty} a_j z^j .$$

Assume that the series in the right-hand side does not reduce to a polynomial and that its radius of convergence is $\sigma_0 (0 < \sigma_0 \leq +\infty)$.

The study of the distribution of the zeros of the partial sums (or sections)

$$(1.2) \qquad s_m(z) = \sum_{j=0}^{m} a_j z^j \qquad (m = 1,2,3,\ldots)$$

of the series in (1.1) was initiated by Jentzsch [16] who proved a fundamental

Theorem. If $0 < \sigma_0 < +\infty$, and if $\varepsilon > 0$ and $\phi (0 \leq \phi < 2\pi)$ are given, there are infinitely many sections $s_m(z)$ having at least one zero in the disk

$$|z - \sigma_0 e^{i\phi}| \leq \varepsilon \quad .$$

Shortly after Jentzsch's discovery, Szegö [33] proved that the conditions of Jentzsch's theorem imply the existence of an infinite sequence of positive increasing integers such that as $m \to +\infty$, by values of this sequence, the arguments of the zeros of $s_m(z)$ are equidistributed in the sense of Weyl.

It is clear that if $\sigma_0 = +\infty$, that is if $f(z)$ is entire, the statements of Jentzsch and Szegö must be modified. The problem of finding substitutes for the Jentzsch-Szegö theorem, applicable to entire functions, is not new and, in first approximation, it is completely solved by the classical results of Carlson [3], [4], and Rosenbloom [26], [27]. Carlson stated his results in 1924 (without proofs) and published his proofs in 1948. The first published proofs were presented in 1943, by Rosenbloom, in his remarkable thesis (Stanford, 1943).

A major contribution to the subject is due to Szegö [34]. In this paper, which preceded the statement of Carlson's results, Szegö undertook a penetrating study of the zeros of the sections $s_m(z)$ of the Taylor expansion of $\exp(z)$. His remarkable analysis may well have prompted all the subsequent developments of the theory.

* * * *

From this point on, we always assume that $f(z)$, in (1.1), is an entire function of order λ :

$$0 < \lambda < +\infty .$$

In spite of their interest, we do not deal with the limiting cases $\lambda = 0$ and $\lambda = +\infty$, because it will soon appear, from the nature of our proofs, that they require a separate (probably lengthy) treatment.

From (1.1) and (1.2) it follows that, as $m \to +\infty$, the partial sums $s_m(z)$ converge uniformly to $f(z)$, on every compact set of the z-plane. Hence, if m is large, some zeros of $s_m(z)$ will be close to some zeros of $f(z)$, and it is natural to expect that many other zeros of the partial sums will be unaccounted for. These "spurious" zeros are a manifestation of the truncation process which defines $s_m(z)$, and one of our goals is to discover some pattern in their behavior.

Szegö's choice of $f(z) = \exp(z)$ as the subject of his investigation is particularly fortunate, not only because of the intrinsic importance of the exponential, but also because (since $\exp(z)$ has no zeros) all the zeros of all the associated sections $s_m(z)$ are spurious. As interesting as they may be, Szegö's results do not reveal the influence, on the zeros of $s_m(z)$, of the order of the function under consideration.

This remark has prompted us to take, in (1.1),

$$a_j = \frac{1}{\Gamma\left(1 + \frac{j}{\lambda}\right)} \qquad (1 < \lambda < +\infty), \tag{1.3}$$

so that, in fact, we shall be studying the expansion

$$f(z) = E_{1/\lambda}(z) = \sum_{j=0}^{\infty} \frac{z^j}{\Gamma\left(1 + \frac{j}{\lambda}\right)}, \tag{1.4}$$

where $E_{1/\lambda}(z)$ is the Mittag-Leffler function of order λ (see, for example, [5]).

Sections 4-16 of this monograph are devoted to the study of the partial sums $s_m(z)$ associated with $E_{1/\lambda}(z)$ $(1 < \lambda < +\infty)$.

In sections 17-25 we examine, from the same point of view, the partial sums associated with a class of entire functions of order $\lambda (0 < \lambda < 1)$. The exact definition of this class will be found on p. 19; its members will be called $\mathcal{L}$-<u>functions of genus zero</u>.

All Mittag-Leffler functions $E_{1/\lambda}(z)$, with $0 < \lambda \leq 1/2$, are $\mathcal{L}$-functions. For sake of brevity, we have omitted the study of the Mittag-Leffler functions of order $\lambda \in (1/2,1)$. They offer no additional difficulties and will be included, as special cases of general classes, in a forthcoming paper of Edrei [9].

The case $\lambda = 1$, which corresponds to $E_1(z) = \exp(z)$, has been so exhaustively investigated by Szegö and others (e.g. Buckholtz [2] and Dieudonné [6]) that its study need not be repeated here. On the other hand, a critical examination of the results regarding exp(z) is important because it invariably suggests conjectures, and sometimes proofs, which cover more general situations. At the suggestion of Varga, K. E. Iverson numerically computed the zeros of the sections $s_m(z)$ $(1 \leq m \leq 23)$ generated by the Taylor expansion of exp(z). Iverson's results [15] seemed to indicate the existence of a parabolic region, symmetric about the positive axis of the complex plane, which is free from zeros of all the $s_m(z)$.

Several authors (cf. Varga [38], Saff and Varga [29], and Newman and Rivlin [21] - [22]) have studied this phenomenon and proved sharp theorems which fully establish the existence of a parabolic region, such as (cf. [29])

(1.5) $\{z = x + iy : |y| \leq 2(x+1)^{1/2},\ x > -1\}$,

which is free of zeros of all partial sums of exp(z). On the other hand, Newman and Rivlin [21] showed that, for any $\varepsilon > 0$ and for any positive constants K and x_0, the closed set

(1.6) $\{z = x + iy : |y| \leq Kx^{(1/2)+\varepsilon},\ x \geq x_0\}$

contains infinitely many zeros of the partial sums of exp(z).

As a result of the above and additional detailed numerical calculations, Saff and Varga have stated

The Saff-Varga Width Conjecture* ([30], [31]). Consider the "parabolic region"

(1.7) $S_0(\tau) = S_0(\tau;K,x_0) = \{z = x+iy: |y| \le Kx^{1-(\tau/2)},\ x \ge x_0\}$,

where $\tau > 0$, and K and x_0 are given positive constants, and consider also the regions $S_\phi(\tau)$, obtained by rotations of $S_0(\tau)$:

(1.7´) $S_\phi(\tau) = S_\phi(\tau;K,x_0) = \{z : ze^{-i\phi} \in S_0(\tau;K,x_0)\}$.

Then, given any entire function f(z) of order $\lambda > \tau$, it is conjectured that there is no $S_\phi(\tau)$ which is devoid of all zeros of all partial sums $s_m(z)$ of the expansion (1.1) of f(z).

We note that the preceding discussion concerning (1.6) gives the validity of the Width Conjecture in the special case $f(z) = \exp(z)$ and $\phi = 0$. Moreover, we remark, in the special case $\tau = 0$, that the validity of the Width Conjecture directly follows from the previously mentioned results of Carlson and Rosenbloom.

The present effort was prompted by an attempt to settle the Saff-Varga Width Conjecture. Although the general Width Conjecture remains open, the theorems of this paper show that the Width Conjecture is valid for each Mittag-Leffler function $f(z) = E_{1/\lambda}(z)$ with $\lambda \in (0,1/2] \cup (1,+\infty)$, in a significantly more precise form. While this point is not covered here, the Width Conjecture is also true for $1/2 < \lambda \le 1$.

In light of our results (stated in the next section), it seems appropriate to suggest an additional

---

*Although this Width Conjecture was originally given for $0 \le \tau \le 2$, the restriction that $\tau \le 2$ does not seem necessary.

Modified Conjecture. Let the function in (1.1) be entire, of order λ $(0 < \lambda < \infty)$. Then, there always exists an infinite sequence M of positive increasing integers m, and a finite number of exceptional arguments

$$\phi_1, \phi_2, \ldots, \phi_q ,$$

having the following properties:

(i) if

$$\phi \not\equiv \phi_j \pmod{2\pi} \qquad (j = 1,2,\ldots,q) ,$$

it is possible to find a positive sequence $\{\rho_m\}$ $(m \in M)$ with ρ_m depending on ϕ such that

(ii) $$\rho_m \to +\infty \quad (m \to +\infty,\ m \in M) ;$$

(iii) for every given ε $(0 < \varepsilon < 1)$, the number of zeros of $s_m(z)$ in the disk

$$|z - \rho_m e^{i\phi}| \le \rho_m m^{-1+\varepsilon} \tag{1.8}$$

tends to $+\infty$ as $m \to +\infty$, $m \in M$.

If, for an entire function $f(z)$ of order λ, the Modified Conjecture is true with

$$\rho_m = O(m^{2/\lambda}) \qquad (m \to \infty,\ \phi \not\equiv \phi_j \pmod{2\pi}) ,$$

then it is immediate that the Width Conjecture is also valid for this function, except, perhaps, along a finite number of rays through the origin.

The theorems of this paper (to be given in §2) show that both the Width Conjecture and the Modified Conjecture are valid for each Mittag-Leffler function $f(z) = E_{1/\lambda}(z)$ with $\lambda \in (0,1/2] \cup (1,+\infty)$, and for each $\mathcal{L}$-function of genus zero (cf. §2, part II). (As mentioned earlier, the interval

(1/2,1] is not exceptional (cf. [9]) so that λ may range over the whole interval $(0, +\infty)$.) As permitted by the Modified Conjecture, the positive axis is an exceptional ray. It requires the replacement of (1.8) by

$$(1.9) \qquad |z - \rho_m| \le \rho_m m^{-(\frac{1}{2}) + \varepsilon} .$$

Some recent (unpublished) work of Edrei suggests that, if the growth of f(z) is sufficiently irregular, further deterioration of (1.8) is possible: it may have to be replaced by

$$(1.10) \qquad |z - \rho_m e^{i\phi}| \le \rho_m m^{-(\frac{1}{k}) + \varepsilon} ,$$

where ϕ is one of the exceptional arguments and $k > 2$ is an integer.

2. Statements of our results. (I). Mittag-Leffler functions of order $\lambda > 1$. In this first portion of our statements, which includes Theorems 1 - 6, we assume that a_j is given by (1.3) with $1 < \lambda < \infty$. Since the order λ remains fixed and is subject to no restrictions other than $\lambda \in (1,\infty)$, these theorems disclose the influence of the order on the nature of the results.

In most of our proofs, and some of our statements of Theorems 1 - 6, we introduce an auxiliary complex variable

$$(2.1) \qquad w = zR^{-1} ,$$

where

$$(2.2) \qquad R = R_m \equiv (m/\lambda)^{1/\lambda}\exp(1/2m) \qquad (1 \le m = \text{integer}).$$

The behavior of the normalized sections

(2.3) $$s_m(R_m w)$$

is easier to grasp than that of the original sections $s_m(z)$. The fact that the normalization in (2.3) is a function of the degree m of the section under consideration is easily understood, and will not create difficulties.

The reader familiar with Wiman-Valiron's theory of the maximum term will notice that $a_m R^m$ is the <u>maximum term</u> of the series in (1.4), and consequently $m = m(R)$ is the <u>central index</u>. These facts should prove essential in the treatment of general entire functions. They are not needed here because a direct study of the situation is simpler and more accurate than the one that we could derive from the Wiman-Valiron theory.

The zeros of $E_{1/\lambda}(z)$ for $1 < \lambda < \infty$ are close to the two rays $\arg z = \pm \pi/2\lambda$. Hence, as a consequence of the uniform convergence

(2.4) $$s_m(z) \to E_{1/\lambda}(z) \quad , \quad (m \to +\infty),$$

on compact subsets, there will be zeros of $s_m(R_m w)$ in the neighborhood of the two rays $\arg w = \pm \pi/2\lambda$. In a sense that we shall make precise, all the other zeros of $s_m(R_m w)$ will be, for large values of m, close to a curve S that we call the <u>normalized Szegö curve</u> of the function under consideration. Moreover, as $m \to +\infty$, every point of S will be a limit point of zeros of the sequence $s_m(R_m w)$.

<u>Definition of S. By $S = S(\lambda)$, we denote a simple closed curve of the w-plane which is represented, in polar coordinates, as the set of all points</u>

(2.5) $$\xi = \xi(\phi) = \sigma(\phi)e^{i\phi} \qquad \left(-\frac{\pi}{2\lambda} \le \phi < 2\pi - \frac{\pi}{2\lambda}\right)$$

where,

(i) <u>for</u> $-\frac{\pi}{2\lambda} \le \phi \le \frac{\pi}{2\lambda}$, $\sigma(\phi)$ <u>is the unique solution of</u>

(2.6) $$\{\sigma(\phi)\}^{\lambda}\cos(\phi\lambda) - 1 - \lambda \log \sigma(\phi) = 0 \qquad (e^{-1/\lambda} \le \sigma(\phi) \le 1);$$

(ii) <u>for</u> $\frac{\pi}{2\lambda} < \phi < 2\pi - \frac{\pi}{2\lambda}$, <u>we take</u>

$$\sigma(\phi) = e^{-1/\lambda} .$$

The fact that there is a unique $\sigma(\phi)$ satisfying the conditions (2.6) is almost obvious (it is established in §10).

It is clear that the normalized Szegö curve is constituted by the juxtaposition of three analytic arcs:

(i) $\xi(\phi) = \sigma(\phi)e^{i\phi} \qquad (0 \le \phi \le \frac{\pi}{2\lambda})$;

(ii) the symmetrical arc

$$\xi(\phi) = \sigma(\phi)e^{i\phi} \qquad \left(-\frac{\pi}{2\lambda} \le \phi \le 0\right)$$

whose points are complex conjugates of the points of the arc in (i);

(iii) a circular arc, with center at the origin, and passing through extremities

(2.7) $$\exp\left(-\frac{1}{\lambda} \pm i\frac{\pi}{2\lambda}\right),$$

of the other two arcs.

The tangent to the Szegö curve is undefined at three points: 1 and the two points in (2.7).

A number of auxiliary quantities attached to S play a role in our statements. Using (2.6), define τ by

(2.8) $$\{\xi(\phi)\}^{\lambda}-1-\lambda\log \xi(\phi) = i(\{\sigma(\phi)\}^{\lambda}\sin(\phi\lambda) - \phi\lambda) = i\tau$$

$$\left(-\frac{\pi}{2\lambda} < \phi < \frac{\pi}{2\lambda}\right).$$

We also note that the tangent to S at $\sigma(\phi)e^{i\phi}$ $(0 < \phi < \frac{\pi}{2\lambda})$ is determined by

$$(2.9) \qquad \frac{d\xi}{d\phi} = i\xi \, \frac{1-(\overline{\xi}(\phi))^{\lambda}}{1-\sigma^{\lambda}\cos(\phi\lambda)} \quad ,$$

and hence, the "outer normal" is given by

$$(2.10) \qquad -i \, \frac{d\xi}{d\phi} = \frac{\xi}{1-\xi^{\lambda}} \cdot \frac{|1-\xi^{\lambda}|^2}{1-\sigma^{\lambda}\cos(\phi\lambda)} \; .$$

We now state the main results of our investigation of Mittag-Leffler functions of order $1 < \lambda < \infty$ as a sequence of four theorems, each of which describes those zeros of $s_m(z)$ which are close to some point

$$R_m\xi(\phi) \qquad (\xi(\phi) \in S) \; .$$

(The totality of all such points will be denoted by $R_m S$.) The picture will be completed by our Theorems 5 and 6 which outline the regions of the z-plane free from zeros of $s_m(z)$.

Theorem 1. *Consider the polynomial*

$$(2.11) \qquad s_m\left(R_m\left(1 + \left(\frac{2}{\lambda m}\right)^{1/2} \zeta\right)\right) = \Phi_m(\zeta),$$

where $s_m(z)$ *is the* m-th *partial sum of* (1.4) (*with* $1 < \lambda < \infty$) *and*

$$\zeta = \rho e^{i\theta} \qquad (\rho \geq 0, \; \theta \text{ real})$$

is an auxiliary complex variable. Let $\operatorname{erfc}(\zeta)$ *denote the complementary error function*

$$(2.12) \qquad \operatorname{erfc}(\zeta) = 1 - \frac{2}{\sqrt{\pi}}\int_0^{\zeta} e^{-v^2}\,dv = \frac{2}{\sqrt{\pi}}\int_{\zeta}^{\infty} e^{-v^2}\,dv \; .$$

Then, as $m \to +\infty$, *we have*

$$(2.13)\quad \left(1+\left(\frac{2}{\lambda m}\right)^{1/2}\zeta\right)^{-m}\{f(R_m)\}^{-1}\, s_m\left(R_m\left(1+\left(\frac{2}{\lambda m}\right)^{1/2}\zeta\right)\right) \rightarrow \frac{1}{2}\exp(\zeta^2)\,\mathrm{erfc}(\zeta)\quad ,$$

uniformly on every compact set of the ζ-plane.

Let $t > 0$ be selected such that on the circle $|\zeta| = t$,

$$\mathrm{erfc}(\zeta) \neq 0 \quad ,$$

(t is otherwise arbitrary) and let

$$(2.14)\qquad \zeta_1, \zeta_2, \ldots, \zeta_\nu \qquad (\nu = \nu(t))$$

denote all the zeros (necessarily simple) of $\mathrm{erfc}(\zeta)$ in the disk $|\zeta| \leq t$. Then, it is possible to find ν sequences $\{\eta_{m,j}\}_m$ ($j = 1,2,\ldots, \nu$) such that

$$(2.15)\qquad \eta_{m,j} \rightarrow 0 \qquad (m \rightarrow +\infty;\ j = 1,2,\ldots, \nu)$$

and such that, for $m > m_0 = m_0(t)$,

(i) the polynomial $\Phi_m(\zeta)$ of (2.11) has, in the disk $|\zeta| \leq t$, simple zeros at

$$(2.16)\qquad \zeta_1 + \eta_{m,1},\ \zeta_2 + \eta_{m,2}, \ldots,\ \zeta_\nu + \eta_{m,\nu};$$

(ii) $\Phi_m(\zeta)$ has no other zeros in the disk $|\zeta| \leq t$.

From the well-known elementary and asymptotic properties of the complementary error function, we deduce (cf. §8) the

Additional Information

All the zeros ζ_j of $\mathrm{erfc}(\zeta)$ satisfy

$$(2.17)\qquad \mathrm{Re}(\zeta_j) < 0 \ .$$

With $\nu(t)$ defined as in (2.14),

$$(2.18)\qquad \nu(t) \sim \frac{t^2}{\pi} \qquad (t \rightarrow \infty).$$

For those zeros ζ_j of $\mathrm{erfc}(\zeta)$ in the half-plane $\mathrm{Im}(\zeta) \geq 0$, we have

(2.19) $$\arg \zeta_j \to \frac{3\pi}{4} \quad \left(|\zeta_j| \to +\infty\right).$$

Our next theorem is concerned with the behavior of $s_m(z)$ near the image

$$R_m\xi(\phi) \qquad (0 < \phi < \frac{\pi}{2\lambda})$$

in the z-plane of the point $\xi(\phi) \in S$.

Theorem 2. Let $\xi = \xi(\phi)$ $(0 < \phi \leq \frac{\pi}{2\lambda})$ be a fixed point of the Szegö curve S. If $\phi = \frac{\pi}{2\lambda}$, assume in addition that

(2.20) $$\lambda < 2(1+e^2).$$

Let (cf. (2.8))

$$\tau = |\xi|^\lambda \sin(\phi\lambda) - \phi\lambda ,$$

and let the sequence $\{\tau_m\}_m$ be defined by the conditions

(2.21) $$\frac{\tau}{\lambda} m \equiv \tau_m \pmod{2\pi}, \qquad -\pi < \tau_m \leq \pi.$$

Let

(2.22) $$\zeta_0 = -\frac{1}{2}\log(2\pi\lambda) + \frac{1}{2}\left(1-\xi^\lambda\right) + \log\left(\frac{\xi}{1-\xi}\right),$$

where $\log(2\pi\lambda)$ is real and the determination of the last logarithm is such that

(2.23) $$-\pi < \mathrm{Im}\log\left(\frac{\xi}{1-\xi}\right) \leq \pi.$$

Put

(2.24) $$p_m = \left(1 + \frac{\log m + 2i\tau_m - 2\zeta_0}{2(1-\xi^\lambda)m}\right) R_m\xi$$

and consider all the zeros of the polynomial in ζ:

$$(2.25)\qquad s_m\left(p_m - \frac{R_m\xi}{(1-\xi^\lambda)m}\zeta\right) = \psi_m(\zeta) \quad ,$$

where $s_m(z)$ is the m-th partial sum of (1.4) (with $1 < \lambda < \infty$).

A. Then, given t ($t > 0$, $t \neq$ integral multiple of 2π), the polynomial $\psi_m(\zeta)$ ($m > m_0(t)$) has, in the disk $|\zeta| \leq t$, exactly

$$(2.26)\qquad 2\left[\frac{t}{2\pi}\right] + 1 = 2\ell + 1$$

zeros, all of them simple. (As usual, $[x]$ denotes here the largest integer $\leq x$.) Denoting those zeros of $\psi_m(\zeta)$ by $\zeta_{m,k}$ ($k = 0, \pm 1, \ldots, \pm \ell$), then

$$(2.27)\qquad \zeta_{m,k} = 2k\pi i + \eta_{m,k} \qquad (k = 0, \pm 1, \ldots, \pm \ell),$$

where for fixed t,

$$(2.28)\qquad \lim_{m\to\infty} \eta_{m,k} = 0 \qquad (k = 0, \pm 1, \ldots, \pm \ell).$$

In view of (2.5), we see that (2.24), (2.25), and (2.27) imply the following behavior of the zeros of $s_m(z)$ in $|\arg z| < \pi/2\lambda$: these zeros are very regularly distributed on an arc "parallel" to the Szegö curve $R_m S$ and lying outside the bounded region enclosed by $R_m S$. It is remarkable that the distance of the zeros to the Szegö curve $R_m S$ is of the order $(\log m)/m^{1-(1/\lambda)}$, whereas the distance between neighboring zeros of $s_m(z)$ is of the order $1/m^{1-(1/\lambda)}$.

For points near the circular portion of the Szegö curve we prove a result entirely analogous to Theorem 2. The only difference between Theorem 2 and the following Theorem 3 is due to the fact that the relevant parameters have different values.

Theorem 3. Let

$$(2.29) \qquad \xi = e^{-\frac{1}{\lambda}} e^{i\phi} \qquad \frac{\pi}{2\lambda} < \phi \le \pi$$

be a point of the circular portion of S, and let the sequence $\{\tilde{\tau}_m\}_m$ be defined by the conditions

$$(2.30) \qquad \tilde{\tau}_m \equiv (m+1)\phi \qquad (\mathrm{mod}\ 2\pi), \qquad -\pi < \tilde{\tau}_m \le \pi.$$

Let

$$(2.31) \qquad \tilde{\zeta}_0 = \frac{1}{2}\log(2\pi) + \left(\frac{1}{\lambda} - \frac{1}{2}\right)(1+\log\lambda) - \log\Gamma\left(1 - \frac{1}{\lambda}\right) + \log\left(e^{1/\lambda}e^{-i\phi} - 1\right) + i\pi,$$

where the determinations of the logarithms are chosen such that

$$-\pi < \mathrm{Im}(\tilde{\zeta}_0) \le \pi.$$

Put

$$(2.32) \qquad \tilde{p}_m = R_m \exp\left(-\frac{1}{\lambda} + i\phi + \left(\frac{1}{2} - \frac{1}{\lambda}\right)\frac{\log m}{m+1} + \frac{\tilde{\zeta}_0 - i\tilde{\tau}_m}{m+1}\right)$$

and consider all the zeros of the polynomial

$$(2.33) \qquad s_m\left(\tilde{p}_m\left(1 + \frac{\zeta}{m+1}\right)\right) = \tilde{\psi}_m(\zeta),$$

where $s_m(z)$ is the m-th partial sum of (1.4) (with $1 < \lambda < \infty$). Then, denoting by $\tilde{\zeta}_{m,k}$ the zeros of $\tilde{\psi}_m(\zeta)$, the statement A of Theorem 2 holds with $\psi_m(\zeta)$, $\zeta_{m,k}$, $\eta_{m,k}$ replaced, respectively, by $\tilde{\psi}_m(\zeta)$, $\tilde{\zeta}_{m,k}$, $\tilde{\eta}_{m,k}$.

In our next result, we consider the limiting case of Theorem 3 characterized by $\phi = \pi/2\lambda$.

Theorem 4. Consider the angular point

$$(2.34) \qquad \xi = e^{-\frac{1}{\lambda} + \frac{i\pi}{2\lambda}}$$

of S, and let the real quantity α be such that

(2.35) $\alpha > \frac{e}{\lambda}$.

Define the sequence $\{\tau'_m\}_m$ by the conditions

(2.36) $\tau'_m \equiv (m+1)\frac{\pi}{2\lambda} + \alpha \log m \pmod{2\pi}, \; -\pi < \tau'_m \leq \pi.$

Let

(2.37)
$$\zeta'_0 = \frac{1}{2}\log(2\pi) + \left(\frac{1}{\lambda} - \frac{1}{2}\right)\log(e\lambda) - \log\Gamma\left(1 - \frac{1}{\lambda}\right) + \log\left(e^{1/\lambda}e^{-\frac{i\pi}{2\lambda}} - 1\right) + i\pi ,$$

where the determinations of the logarithms are chosen such that

$$-\pi < \mathrm{Im}(\zeta'_0) \leq \pi$$

Put

(2.38)
$$p'_m = R_m \exp\left(-\frac{1}{\lambda} + i\frac{\pi}{2\lambda} + \left\{\left(\frac{1}{2} - \frac{1}{\lambda}\right) + i\alpha\right\}\frac{\log m}{m+1} + \frac{\zeta'_0 + i\tau'_m}{m+1}\right),$$

and consider all the zeros of the polynomial

(2.39)
$$s_m\left(p'_m\left(1 + \frac{\zeta}{m+1}\right)\right) = \chi_m(\zeta) ,$$

where $s_m(z)$ is the m-th partial sum of (1.4) (with $1 < \lambda < \infty$). Then, denoting by $\zeta'_{m,k}$ the zeros of $\chi_m(\zeta)$, the statement A of Theorem 2 holds with $\psi_m(\zeta)$, $\zeta_{m,k}$, $\eta_{m,k}$ replaced, respectively, by $\chi_m(\zeta)$, $\zeta'_{m,k}$, $\eta'_{m,k}$.

It is obvious that we could also formulate a modified form of Theorem 4, valid when the condition (2.35) is replaced by

$$\frac{e}{\lambda} > \alpha .$$

For sake of brevity, we shall not examine this point; its main interest would be to ascertain the "best possible" character of Theorem 4.

In contrast with (2.13) of Theorem 1, we have not explicitly stated in Theorems 2 - 4 the limit functions of our transformations of the $s_m(z)$. In the special case of $\mathcal{L}$-functions, the limit functions in question are completely described by assertion IV of Theorem 7. The corresponding assertions, that we could add to Theorems 2 - 4, have been omitted because they are not very simple, and shed no additional light on the position of the zeros of the partial sums (which is our main interest).

Our next result (Theorem 5 below) shows that if m is large enough, the partial sum $s_m(z)$ has no zeros other than those which lie in the vicinity of the Szegö curve $R_m S$ and of the two rays $\arg z = \pm \frac{\pi}{2\lambda}$.

Theorem 5. Let $s_m(z)$ be the m-th partial sum of (1.4) (with $1 < \lambda < \infty$) . Then,

I. For all m large enough, $s_m(z)$ has no zeros in the annulus

$$|z| \geq R_m \qquad \left(R_m = \left(\frac{m}{\lambda}\right)^{1/\lambda} e^{\frac{1}{2m}}\right) .$$

II. If h is given $(0 < h < \frac{1}{2})$, then $s_m(z)$ $(m > m_0(h))$ has no zeros in each of the sets

$$(2.40) \qquad \left\{z = |z|e^{i\phi}:\ (1+h)R_m\sigma(\phi) \leq |z| \leq R_m;\ |\phi| \leq \pi/2\lambda\right\} ,$$

$$(2.41)\quad \left\{z = |z|e^{i\phi}:\ (1+h)R_m e^{-1/\lambda} \le |z| \le R_m;\ \frac{\pi}{2\lambda} + h \le \phi \le 2\pi - \left(\frac{\pi}{2\lambda} + h\right)\right\},\ (\lambda \ge 2),$$

$$(2.42)\quad \left\{z = |z|e^{i\phi}:\ R_m e^{-1/\lambda} \le |z| \le R_m;\ \frac{\pi}{2\lambda} + h \le \phi \le 2\pi - \left(\frac{\pi}{2\lambda} + h\right)\right\},\ (1 < \lambda < 2),$$

where $\sigma(\phi)$ is defined in (2.6).

III. It is possible to determine a constant $B(h) > 0$, such that

$$(2.43)\quad E_{1/\lambda}(B(h)e^{i\theta}) \neq 0 \qquad (0 \le \theta \le 2\pi)$$

and such that

(i) $s_m(z)$ $(m > m_0(h))$ has in the disk

$$|z| \le B(h)$$

exactly as many zeros as $E_{1/\lambda}(z)$;

(ii) $s_m(z)$ $(m > m_0(h))$ has no zeros in each of the sets

$$(2.44)\quad \left\{z = |z|e^{i\phi}:\ B(h) \le |z| \le R_m\sigma(\phi);\ |\phi| \le \pi/2\lambda\right\},$$

$$(2.45)\quad \left\{z = |z|e^{i\phi}:\ B(h) \le |z| \le R_m e^{-1/\lambda};\ \frac{\pi}{2\lambda} + h \le \phi \le 2\pi - \left(\frac{\pi}{2\lambda} + h\right)\right\}\ (\lambda \ge 2),$$

$$(2.46)\quad \left\{z = |z|e^{i\phi}:\ B(h) \le |z| \le (1-h)R_m e^{-1/\lambda};\ \frac{\pi}{2\lambda} + h \le \phi \le 2\pi - \left(\frac{\pi}{2\lambda} + h\right)\right\},\ (1 < \lambda < 2).$$

It is possible to prove more precise results having the same general character as Theorem 5. This is what is done in the next result (Theorem 6), which studies in detail the vicinity of the vertex $w = 1$ of the normalized Szegö curve.

Theorem 6. *Let ζ^* be the zero of the complementary error function $\operatorname{erfc}(\zeta)$ which has the smallest positive imaginary part, and let K be any constant such that*

$$(2.47) \qquad 0 < K < \frac{\sqrt{2}}{\lambda}\, \operatorname{Im}(\zeta^*).$$

Then, there exists a positive constant $x_0 = x_0(\lambda,K)$ *such that every partial sum* $s_m(z)$ *of* (1.4) (*with* $1 < \lambda < \infty$) *is zero-free in the "parabolic region"*

$$(2.48) \qquad \left\{z = x+iy:\ |y| \le Kx^{1-(\lambda/2)},\ x \ge x_0\right\}.$$

The result is best possible in the sense that there exists a sequence $z_m = x_m+iy_m$, *with* $x_m \to +\infty$, *as* $m \to \infty$, *such that* $s_m(z_m) = 0$ *for* $m > m_0$ *and*

$$(2.49) \qquad \lim_{m \to \infty} \frac{|y_m|}{x_m^{1-(\lambda/2)}} = \frac{\sqrt{2}}{\lambda}\ \operatorname{Im}(\zeta^*).$$

We remark (see [11]) that

$$(2.50) \qquad \zeta^* \doteq -1.354810 + i(1.991467) \quad .$$

Theorems 2 - 4 imply that the Modified Conjecture is true, with $\phi = 0$ as the only exceptional argument, for all Mittag-Leffler functions $E_{1/\lambda}(z)$ with $1 < \lambda < \infty$. Indeed, on taking $\rho_m = R_m\sigma(\phi)$ (where $\xi(\phi) = \sigma(\phi)e^{i\phi}$ is a point of the Szegö curve S) , then for each ϕ with $0 < \phi < 2\pi$ and for each $\varepsilon > 0$, it is easy to see that there exists an infinite sequence $\mathcal{M}$ of positive integers m such that the number of zeros of $s_m(z)$ in the disk

$$(2.51) \qquad |z - \rho_m e^{i\phi}| \leq \rho_m m^{-1+\varepsilon} ,$$

tends to $+\infty$ as $m \to \infty$, $m \in \mathcal{M}$. Moreover, as the above choices for ρ_m satisfy $\rho_m = O(m^{2/\lambda})$, then, as stated in §1, the truth of the Width Conjecture for $0 < \phi < 2\pi$ follows from the truth of the Modified Conjecture. When $\phi = 0$, Theorem 1 implies that the disks

$$(2.52) \qquad |z - R_m| \leq R_m m^{-(1/2)+\varepsilon} \quad (\varepsilon > 0) ,$$

contain many zeros of the $s_m(z)$ for $m \in \mathcal{M}$, where $\mathcal{M}$ is a suitably chosen set of positive integers. Since $R_m = O(m^{1/\lambda})$, it similarly follows that the Width Conjecture is valid when $\phi = 0$. Consequently, for all arguments ϕ, the Width Conjecture is true for all Mittag-Leffler functions $E_{1/\lambda}(z)$ with $1 < \lambda < \infty$.

As a final remark, Theorem 6 shows the Width Conjecture is sharp, in the sense that on letting τ equal λ, there exist positive constants K and x_0 such that $S_0(\lambda;K,x_0)$ is free of zeros of all partial sums $\{s_m(z)\}_{m=1}^{\infty}$, for every Mittag-Leffler function $f(z) = E_{1/\lambda}(z)$ with $1 < \lambda < \infty$.

(II) Functions of genus zero whose zeros are all negative. We conclude our investigation by studying the partial sums of the expansions of some simple functions other than $E_{1/\lambda}(z)$. We shall treat all functions $F(z)$ characterized by the following two conditions.

A. <u>The functions $F(z)$ are entire of order</u> λ $(0 < \lambda < 1)$ <u>and all their zeros are real and negative</u>:

$$F(z) = F(0) \prod_{k=1}^{\infty} \left(1 + \frac{z}{x_k}\right) = \sum_{j=0}^{\infty} a_j z^j \tag{2.53}$$

$$\left(0 < x_k, \sum_{k=1}^{\infty} x_k^{-1} < +\infty, F(0) > 0\right) .$$

B. <u>Along the positive axis</u>

$$\log F(r) = \log M(r) \sim B_1 r^{\lambda} \quad \left(M(r) = \max_{|z|=r} |F(z)|, r \to +\infty, B_1 > 0\right) . \tag{2.54}$$

For ease of reference we shall say that such functions are <u>$\mathcal{L}$-functions of genus zero</u> or, for brevity, simply <u>$\mathcal{L}$-functions</u>.

Our proofs would require minor modifications if we were to replace in (2.54), $B_1 r^{\lambda}$ by $B_1 r^{\lambda} (\log r)^{\alpha}$ (α real) and, more generally, by

$$B_1 r^{\lambda} (\log r)^{\alpha_1} (\log_2 r)^{\alpha_2} \ldots (\log_k r)^{\alpha_k}$$

where the α's are real and $\log_j r$ denotes the iterated logarithm: $\log_j r = \log(\log_{j-1} r)$. Since these generalizations are straightforward, and their treatment does not require new ideas, we shall be content with the simple form in (2.54).

$\mathcal{L}$-functions were systematically studied by Lindelöf [18] and somewhat later by Valiron [37].

Among the well-known $\mathcal{L}$-functions we mention

$$F(z) = \prod_{n=1}^{\infty} \left(1 + \frac{z}{n^{1/\lambda}}\right) \qquad (0 < \lambda < 1) , \tag{2.55}$$

as well as

$$E_{1/\lambda}(z) \qquad (0 < \lambda \leq \tfrac{1}{2}) . \tag{2.56}$$

In §18 we show that (2.55) and (2.56) do in fact define $\mathcal{L}$-functions. We also show that the property of being an $\mathcal{L}$-function of genus zero is invariant under differentiation.

In spite of their very special character, $\mathcal{L}$-functions of genus zero have played a major role in the development of the theory of entire and meromorphic functions. They led Littlewood to conjecture, and Wiman and Valiron to prove, the fundamental $\cos \pi\rho$-theorem. In Nevanlinna's theory, they provided examples of the behavior of deficient values [20] and eventually led to the proof of sharp deficiency relations [10], [7].

Finally, by studying and generalizing the relations (2.53) and (2.54), Valiron discovered an important class of tauberian theorems. His pioneering work was published in 1913 [37], long before Hardy and Littlewood [13] obtained similar results (1930).

We prove

Theorem 7. Let $F(z)$ *be an* $\mathcal{L}$*-function of genus zero and order* $\lambda (0 < \lambda < 1)$ *and let* $s_m(z)$ *be the* m-th *partial sum of its Taylor expansion about* $z = 0$. *Then*

I. *The function*

$$a(r) = r \frac{F'(r)}{F(r)} ,$$

of the positive variable r , *is positive, continuous, strictly increasing and unbounded.*

II. *Define the sequence* $\{R_m\}_m$ *by the conditions*

$$a(R_m) = m \qquad (m = 1,2,3,\ldots) . \tag{2.57}$$

Then, if ζ *is an auxiliary complex variable, we have*

$$(2.58)\quad \left(1+\left(\frac{2}{\lambda m}\right)^{1/2}\zeta\right)^{-m}\left\{F(R_m)\right\}^{-1} s_m\left(R_m\left(1+\left(\frac{2}{\lambda m}\right)^{1/2}\zeta\right)\right) \to \frac{1}{2}\exp(\zeta^2)\,\mathrm{erfc}(\zeta)\ ,$$

uniformly on every compact set of the ζ-plane.

III. With every given ϕ $(0 < |\phi| < \pi)$ it is possible to associate a real sequence $\{\sigma_m(\phi)\}_m$ such that

$$\text{(i)}\qquad \lim_{m\to\infty} \sigma_m(\phi) = \sigma(\phi)\ ,$$

where $\sigma = \sigma(\phi)$ is the unique solution in $(0,1)$ of the equation

$$(2.59)\qquad \sigma^{\lambda}\cos(\phi\lambda) - 1 - \lambda\log\sigma = 0\ ;$$

(ii) write

$$(2.60)\quad \xi_m = \xi_m(\phi) = \sigma_m(\phi)e^{i\phi},\quad \xi = \sigma(\phi)e^{i\phi},\quad L_m = (2\pi\lambda m)^{1/2}\xi_m^{-m}\{F(R_m)\}^{-1}\ ;$$

then the polynomials in ζ

$$(2.61)\qquad L_m s_m\left(R_m\xi_m\left(1+\frac{\zeta}{m(1-\xi^{\lambda})}\right)\right)$$

are uniformly bounded on every compact set of the ζ-plane.

IV. Every limit function of the polynomials in (2.61) is of the form

$$(2.62)\qquad \exp\left(\frac{\zeta}{1-\xi^{\lambda}}\right)\left\{e^{i\chi}\,e^{-\zeta} - \frac{\xi}{1-\xi}\right\} = Z_{\chi}(\zeta)\ ,$$

where the real quantity χ may depend on the particular sequence of integers through which $m \to +\infty$.

The imprecision of the asymptotic relation (2.54) does not permit us to give approximations of R_m and $\sigma_m(\phi)$ as good as the corresponding ones obtained in our study of $E_{1/\lambda}(z)$ $(\lambda > 1)$.

Nevertheless, once the centers R_m or $R_m\sigma_m(\phi)e^{i\phi}$ of the critical disks are taken for granted the behavior of the zeros of $s_m(z)$ in

$$(2.63) \qquad |z - R_m| \le \left(\frac{2}{\lambda m}\right)^{1/2} R_m t \quad ,$$

or in

$$(2.64) \qquad |z - R_m\xi_m(\phi)| \le \frac{R_m|\xi_m(\phi)|}{m|1-\xi^\lambda|}\, t \qquad (0 < |\phi| < \pi) \quad ,$$

is entirely determined by the zeros, in the disk $|\zeta| \le t$, of the corresponding limit functions $\mathrm{erfc}(\zeta)$ and $Z_\chi(\zeta)$.

For t and m large (t fixed), there will be approximately t^2/π zeros of $s_m(z)$ in the disk (2.63) and approximately t/π zeros in the disk (2.64).

It is easily seen, just as in part 2.(I), that the Modified Conjecture and the Width Conjecture are valid for all $\mathcal{L}$-functions of genus zero. (We have not discussed the zeros of the partial sums near the negative real axis. But as $F(z)$ has infinitely many zeros on the ray $(-\infty, 0)$, then any "parabolic region" symmetric about this ray will necessarily contain infinitely many zeros of the partial sums.)

(III) <u>Problems for further study</u>. The preceding investigation displays the influence, for all positive values of the order λ of $f(z)$, on the precise statements which were obtained. It may be of interest to point out that this influence is quite weak: in much of our work, λ is a mere parameter in the changes of variable leading to the limit functions which characterize the local behavior of $s_m(z)$. In the most delicate of our results (Theorem 1 and Theorem 7 (II)), the limit function is independent of λ.

The power series that we have treated, that is the expansions of $E_{1/\lambda}(z)$ and of $\mathcal{L}$-functions of genus zero, are admittedly very special but it is not unlikely that they exhibit a behavior typical of much more general ones.

Among the questions which now seem ready for further study, a few will be singled out.

A. The conjectures. The Saff-Varga Width Conjecture and the Modified Conjecture are to receive serious attention.

B. The limiting cases $\lambda = 0$ and $\lambda = +\infty$. For $\lambda = +\infty$, we may investigate the expansion

$$\sum_{k=2}^{\infty} \frac{z^{k-2}}{\Gamma\left(1 + (\log k)^{-\alpha}\right)} \qquad (0 < \alpha < 1) \quad ,$$

introduced by Malmquist [19] in the same volume of Acta Mathematica which contains Mittag-Leffler's definition and study of $E_{1/\lambda}(z)$.

For $\lambda = 0$, some caution should be exercised because the scattering, in all directions, of the zeros of $s_m(z)$ no longer holds for the partial sums of series representing functions of very slow growth; for instance, as shown by Pólya and Szegö [25] the sections of the expansion

$$\sum_{k=0}^{\infty} 2^{-k^2} z^k$$

have all their zeros negative.

An investigation of functions of order zero should be preceded by a careful study of Ganelius' important paper [12] on the subject.

C. The Padé table. The sequence of partial sums of a power series may be considered as the zero-th row of its Padé table. (In the n-th row we have Padé numerators $P_{mn}(z)$ of degree m and Padé denominators $Q_{mn}(z)$ of degree n.)

How do the numerators $P_{mn}(z)$, of the n-th row, behave? Along these lines there are two patterns to guide us:

(i) the detailed study by Saff and Varga [28], [32] of the Padé approximants of $\exp(z)$;

(ii) Edrei's extension [8] to the complete Padé table of the results of Carlson and Rosenbloom. It should be noted that the extension in question is completely successful and is as precise as the underlying theorems of Carlson-Rosenbloom.

To transfer, to the Padé table, the very precise results of the present monograph may not be easy, but there is no reason to believe that it will present unsurmountable difficulties.

D. Extensions to more general classes of functions. It will be shown by Edrei [9] that those entire functions of order $\lambda (0 < \lambda < +\infty)$, which are also admissible in the sense of Hayman [14], constitute a class of functions to which most of our results are applicable.

Other classes seem ready for an investigation in depth; among the more promising ones we mention the class of functions (introduced and studied, independently, by Pfluger [23] and Levin [17]) whose zeros have an angular density.

3. Discussion of our numerical results. To illustrate our theoretical results of Theorems 1-6 of §2.(I) (Mittag-Leffler functions of order $1 < \lambda < \infty$), we have included here Figures 1-9, which we now explain. First, three distinct values of $\lambda > 1$ were chosen, namely $\lambda_1 = 10/9$, $\lambda_2 = 2$, and $\lambda_3 = 3.5$. The first three figures show all the zeros of the normalized partial sums

$$s_m(R_m z) = s_m(R_m z;\lambda_j) = \sum_{k=0}^{m} \frac{R_m^k z^k}{\Gamma(1+k/\lambda_j)} \qquad (m = 1,2,\ldots,85)$$

of $E_{1/\lambda_j}(z)$, the second three figures show the eighty-five zeros of the normalized partial sums $s_{85}(R_{85}z;\lambda_j)$, while the last three figures show all the zeros of the unnormalized partial sums $\{s_m(z;\lambda_j)\}_{m=1}^{85}$. We remark that each of the first three and last three figures contains 3655 zeros.

First, consider Figures 7-9. From Theorem 6, there exist positive constants K and x_j (j=1,2,3) such that the "parabolic region" (cf. (2.48))

(3.1) $\quad \{z = x+iy: |y| \leq Kx^{1-(\lambda_j/2)},\ x \geq x_j\} \qquad (j=1,2,3)$

contains no zero of any partial sums $s_m(z;\lambda_j)$, $m \geq 1$, for each j=1,2,3. On the other hand, since the Width Conjecture is valid for each Mittag-Leffler function $f(z) = E_{1/\lambda}(z)$, then the closed set

(3.2) $\quad \{z = x+iy: |y| \leq K_j x^{1-(\tau_j/2)},\ x \geq \tilde{x}_j\}$

must, for any choice of $\tau_j (0 < \tau_j < \lambda_j)$ and positive constants K_j and $\tilde{x}_j$, contain infinitely many zeros of the partial sums of $E_{1/\lambda_j}(z)$. Now, because $\lambda_2 = 2$, the set in (3.1) when j = 2 is by definition a semi-infinite strip, symmetric about the positive real axis. That such a strip is the largest such set which is devoid of all zeros of all partial sums of $E_{1/2}(z)$, is particularly evident from the numerical results of Figure 8.

We next remark that eight zeros of $E_{1/\lambda_1}(z)$, four zeros of $E_{1/\lambda_2}(z)$, and two zeros of $E_{1/\lambda_3}(z)$ can be seen, respectively, in Figures 7-9, because of the evident clustering of zeros (Hurwitz' Theorem) of the associated partial sums. The same number of these zeros, respectively, appear very distinctly in the graphing of the zeros of $s_{85}(R_{85}z;\lambda_j)$ in Figures 4-6. In each of Figures 4-6, the solid curve is the associated Szegö curve $S = S(\lambda_j)$, and these zeros of $s_{85}(R_{85}z;\lambda_j)$ are all well interior to $S(\lambda_j)$. Similarly, on examining Figures 1-3, we see these "Hurwitz zeros" of $\{s_m(R_m z;\lambda_j)\}_{m=1}^{85}$ create, by virtue of the normalization, "line segments" of points which are "aimed" at the origin.

Continuing with Figures 4-6, we remark that the dashed lines denote the rays $\phi = \pm \pi/2\lambda_j$. First, we see that in each case, that each zero $\hat{z}$ of $s_{85}(R_{85}z;\lambda_j)$, satisfying $|\arg \hat{z}| < \pi/2\lambda_j$, lies outside the Szegö curve $S(\lambda_j)$, which is consistent with (2.44) of III of Theorem 5. On the other hand, as guaranteed by (2.42) and (2.45) of Theorem 5, given any $h > 0$, there is a positive integer $m_j(h)$ such that in $\pi/2\lambda_j + h \le \arg z \le 2\pi - \pi/2\lambda_j - h$, the zeros of $s_m(R_m z;\lambda_j)$, for any $m \ge m_j(h)$, lie inside (outside) the circular portion of the Szegö curve $S(\lambda_j)$ if $1 < \lambda_j < 2$ (if $\lambda_j \ge 2$). This is again confirmed in Figures 4-6.

Next, a close perusal of (2.33) of Theorem 3 and its associated statement A, gives us that the zeros of $s_m(R_m z;\lambda)$ are asymptotically uniformly distributed in angle, as $m \to \infty$, in the vicinity of the circular portion of the Szegö curve $S(\lambda)$. This too is strikingly evident in Figures 4-6.

Finally, to illustrate our theoretical result of Theorem 7 of §2.(II) ($\mathcal{L}$-functions of genus zero), we have included here Figures 10-12, which we similarly explain. As any Mittag-Leffler function $E_{1/\lambda}(z)$ with $0 < \lambda \leq 1/2$ is an $\mathcal{L}$-function of genus zero (cf. (2.56)), we have chosen the particular value $\lambda = 1/2$. Figure 10 shows all the zeros of the normalized partial sums $s_m(R_m z;\ 1/2)$ $(m = 1,2,\ldots,\ 50)$, where R_m is now defined by (2.57). Figure 11 shows the fifty zeros of the normalized partial sum $s_{50}(R_{50} z;\ 1/2)$. Figure 12 then shows all the zeros of the unnormalized partial sums $\{s_m(z;1/2)\}_{m=1}^{50}$. We remark that each of Figures 10 and 12 contains 1275 zeros.

The solid curve which appears in Figures 10 and 11 is the Szegö curve of (2.59). We see in these figures that the spurious zeros of $s_m(R_m z;\ 1/2)$ lie in a vicinity of this curve. On the other hand, as the zeros of any $\mathcal{L}$-function are all necessarily real and negative (cf. (2.53)), the associated "Hurwitz zeros" of $s_{50}(R_{50} z;\ 1/2)$ now appear on the negative real axis, as Figure 11 clearly shows.

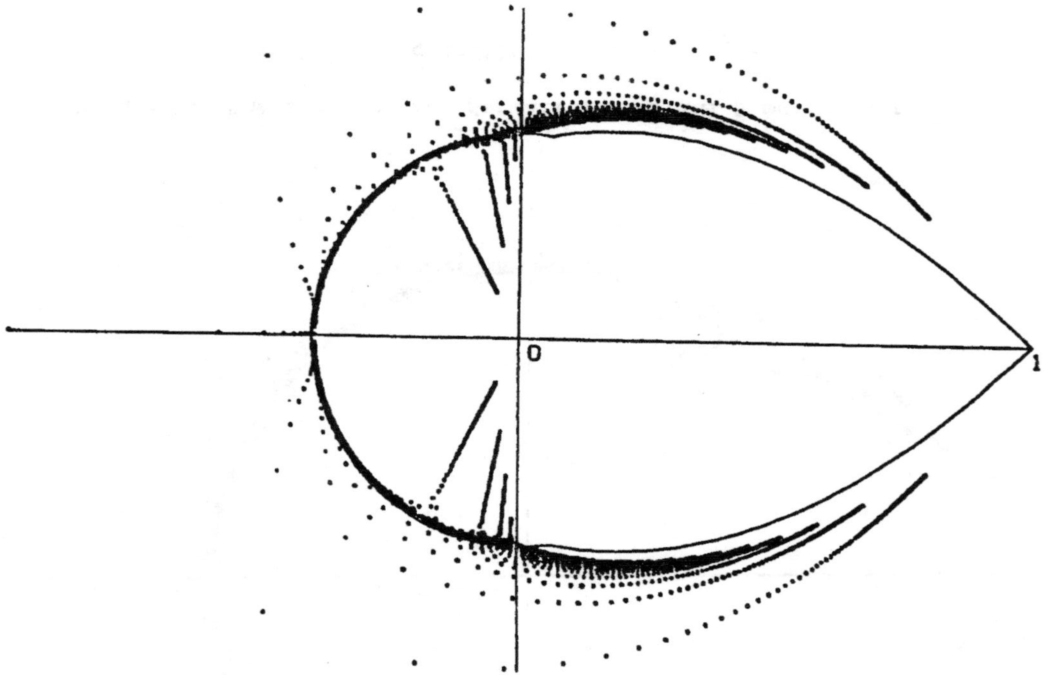

Figure 1. The Zeros of the Normalized Partial Sums $\{s_m(R_m z)\}_{m=1}^{85}$ for the Mittag-Leffler Function $E_{1/\lambda}(z)$ for $\lambda = 10/9$.

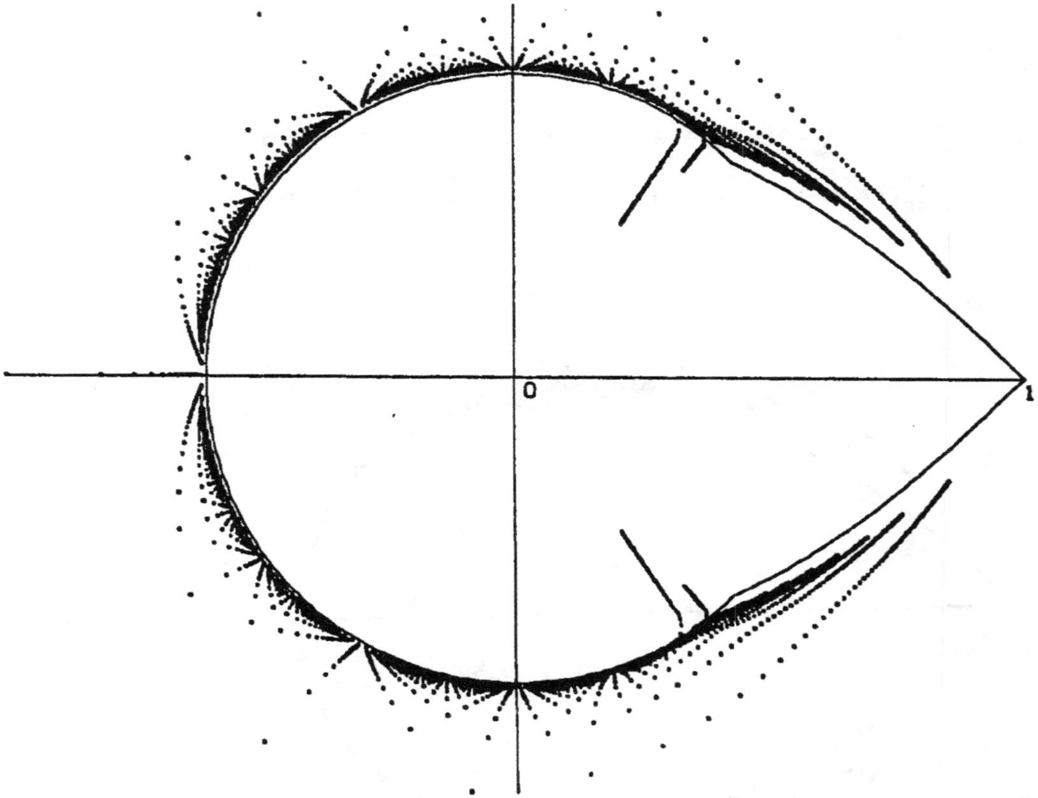

Figure 2. The Zeros of the Normalized Partial Sums $\{s_m(R_m z)\}_{m=1}^{85}$ for the Mittag-Leffler Function $E_{1/\lambda}(z)$ for $\lambda = 2$.

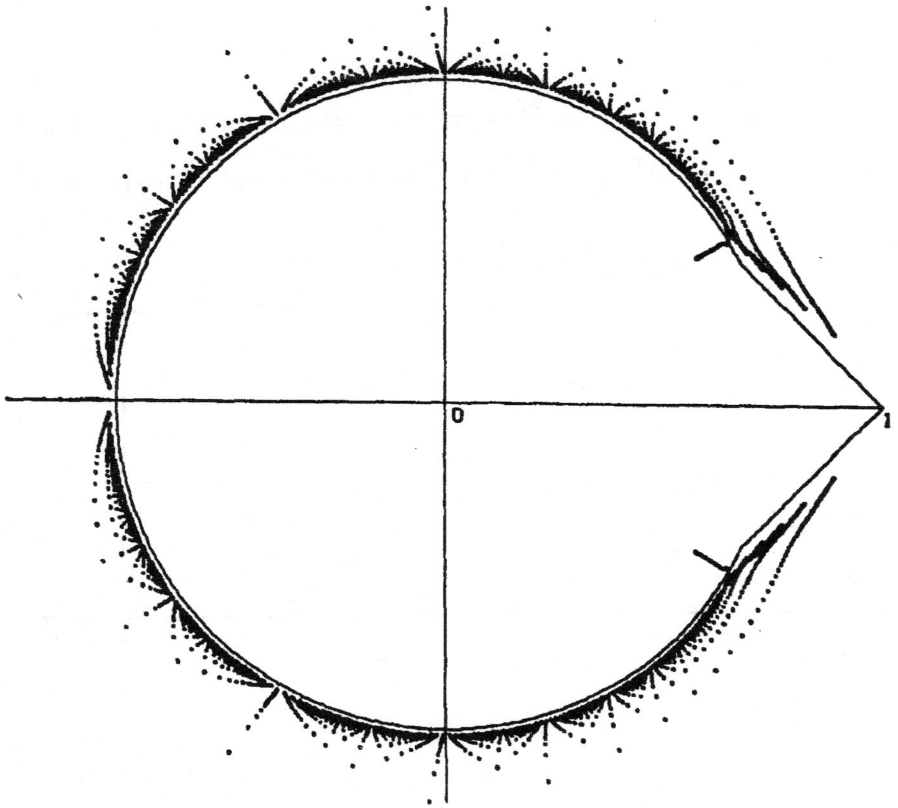

Figure 3. The Zeros of the Normalized Partial Sums $\{s_m(R_m z)\}_{m=1}^{85}$ for the Mittag-Leffler Function $E_{1/\lambda}(z)$ for $\lambda = 3.5$.

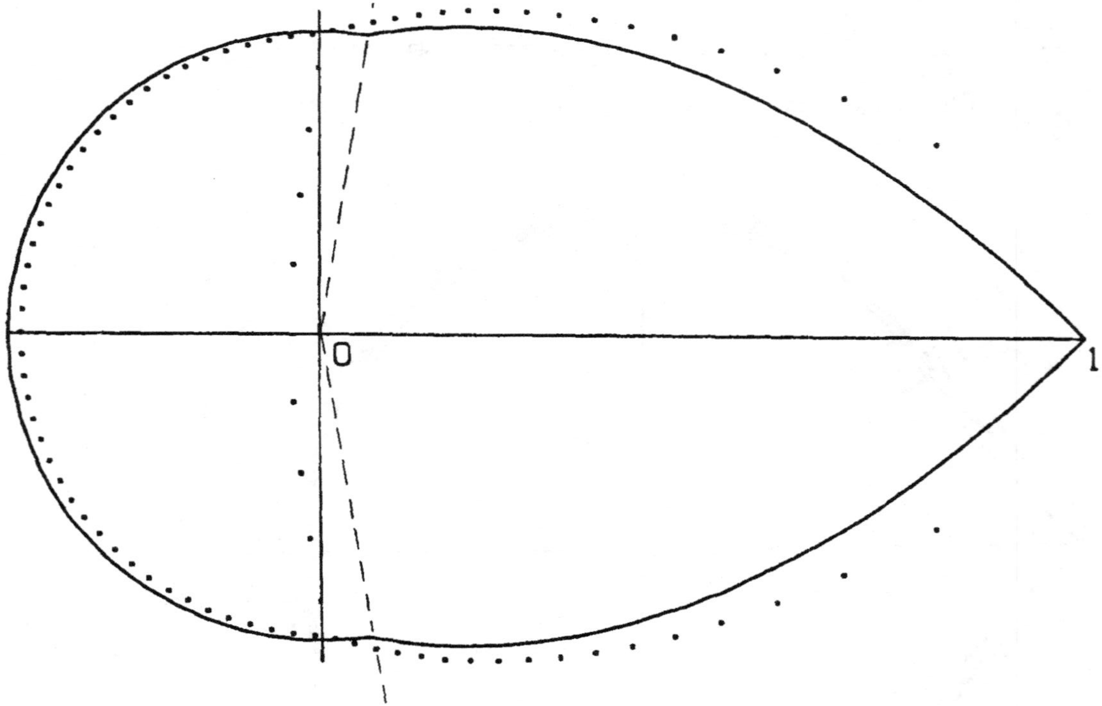

Figure 4. The Zeros of the Normalized Partial Sum $s_{85}(R_{85}z)$ for the Mittag-Leffler Function $E_{1/\lambda}(z)$ for $\lambda = 10/9$.

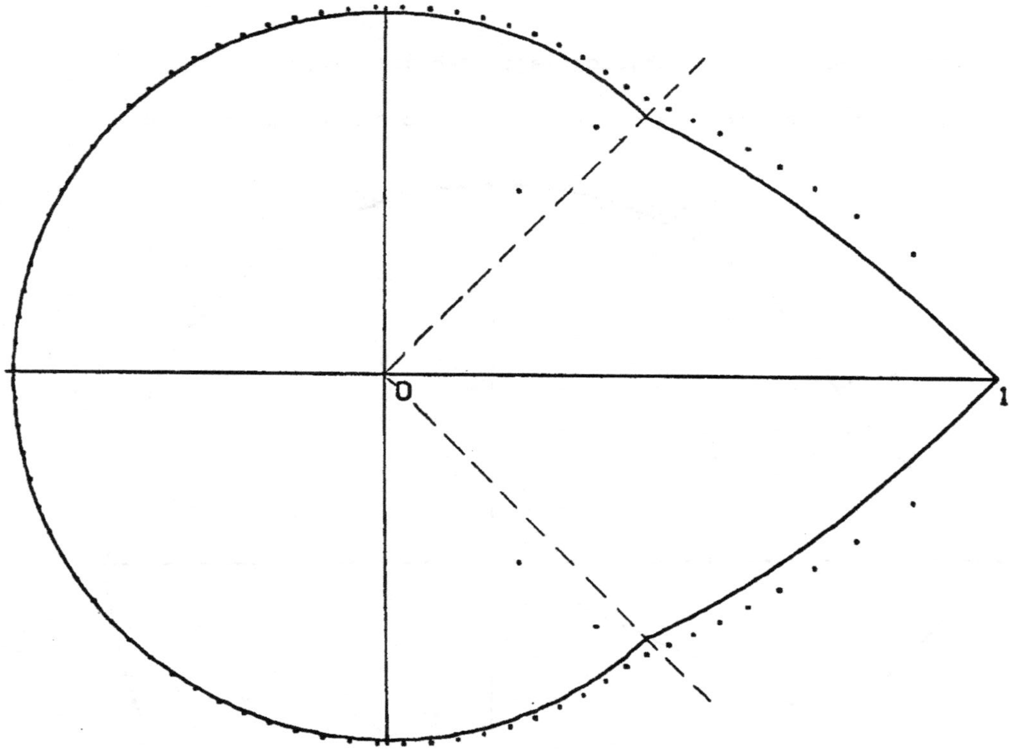

Figure 5. The Zeros of the Normalized Partial Sum $s_{85}(R_{85}z)$ for the Mittag-Leffler Function $E_{1/\lambda}(z)$ for $\lambda = 2$.

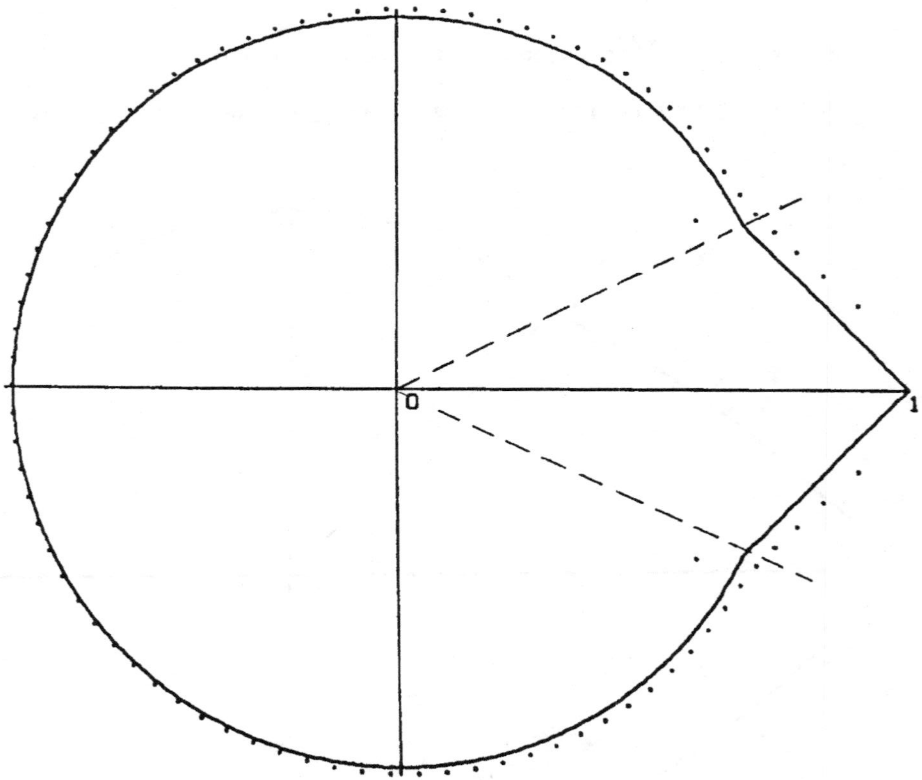

Figure 6. The Zeros of the Normalized Partial Sum $s_{85}(R_{85}z)$ of the Mittag-Leffler Function $E_{1/\lambda}(z)$ for $\lambda = 3.5$.

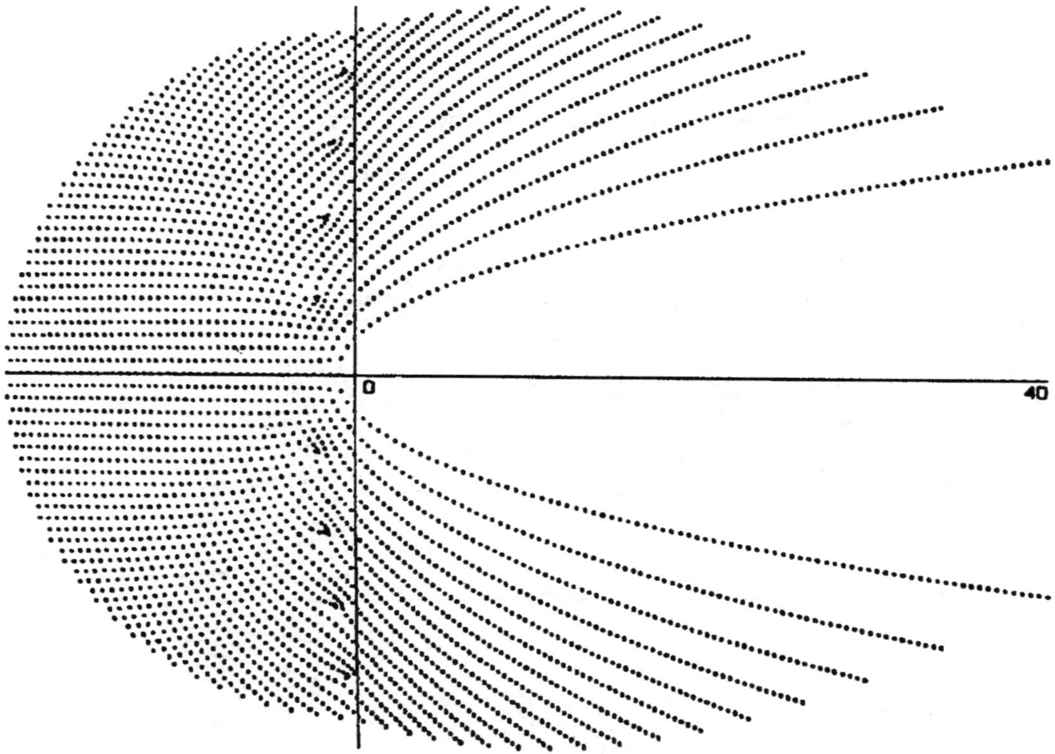

Figure 7. The Zeros of the Partial Sums $\{s_m(z)\}_{m=1}^{85}$ for the Mittag-Leffler Function $E_{1/\lambda}(z)$ for $\lambda = 10/9$.

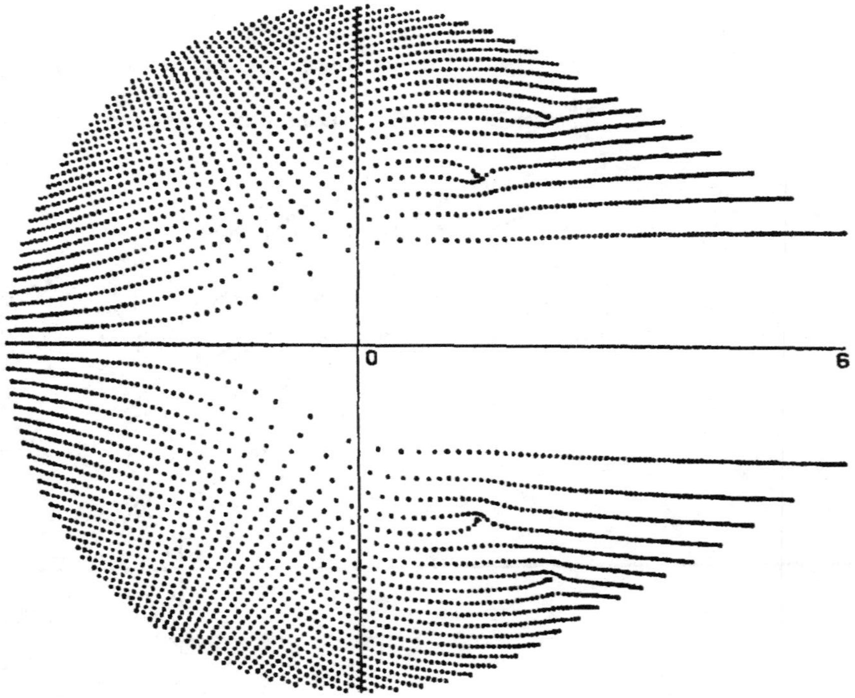

Figure 8. The Zeros of the Partial Sums $\{s_m(z)\}_{m=1}^{85}$ for the Mittag-Leffler Function $E_{1/\lambda}(z)$ for $\lambda = 2$.

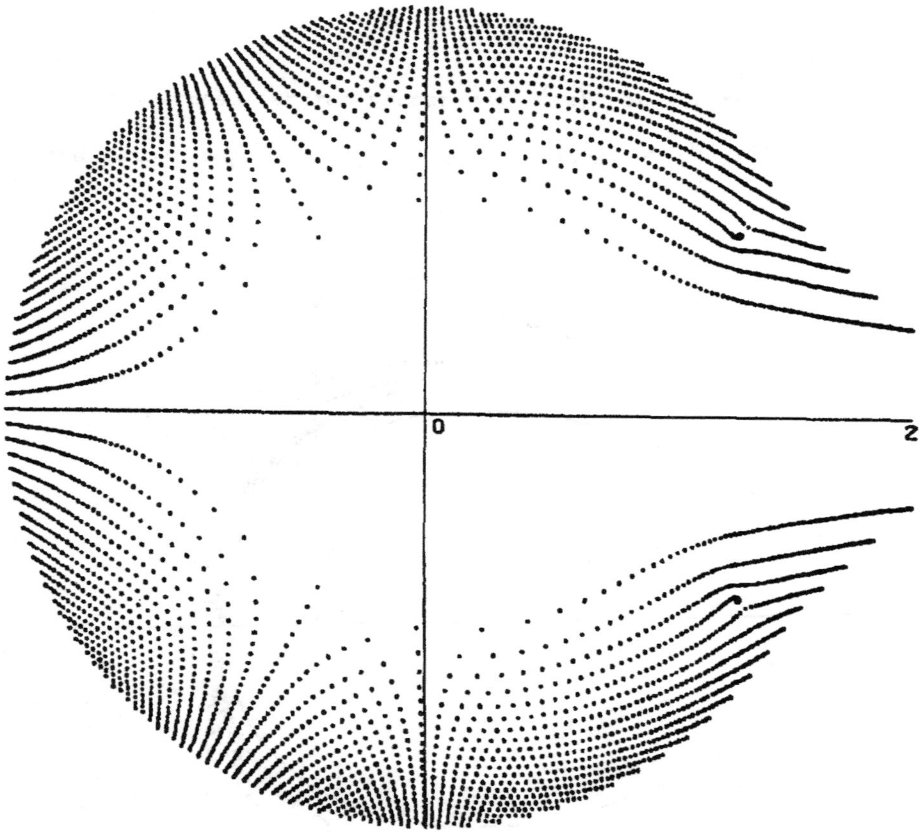

Figure 9. The Zeros of the Partial Sums $\{s_m(z)\}_{m=1}^{85}$ for the Mittag-Leffler Function $E_{1/\lambda}(z)$ for $\lambda = 3.5$.

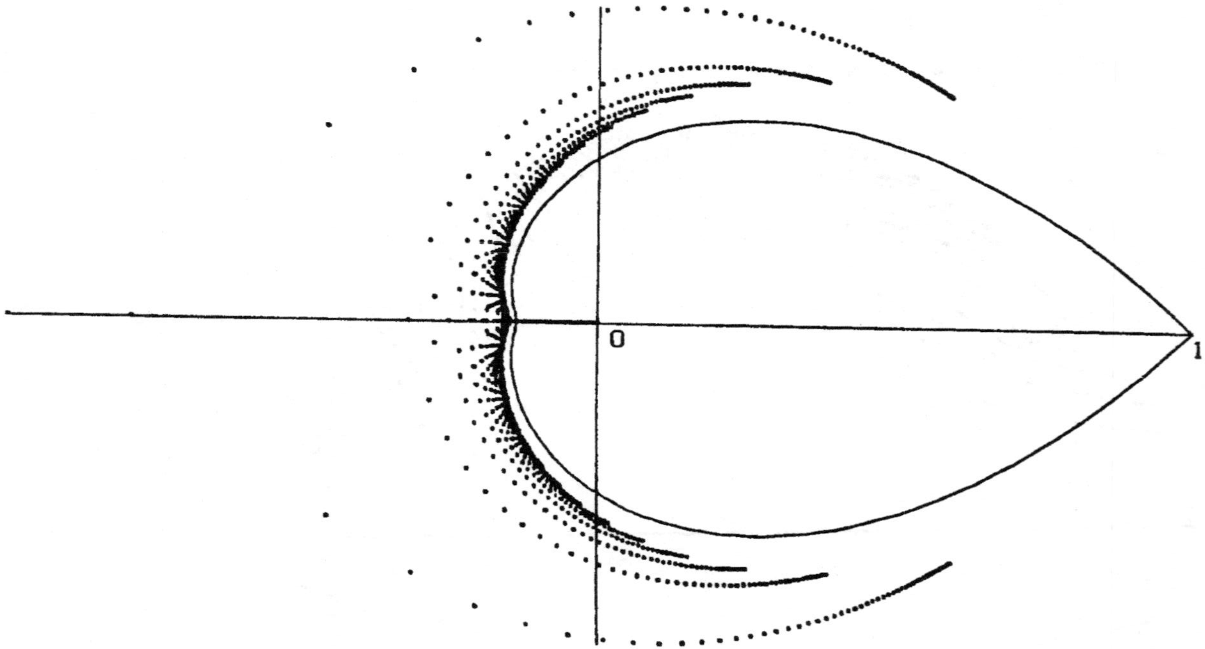

Figure 10. The Zeros of the Normalized Partial Sums $\{s_m(R_m z)\}_{m=1}^{50}$ for the Mittag-Leffler Function $E_{1/\lambda}(z)$ for $\lambda = 1/2$.

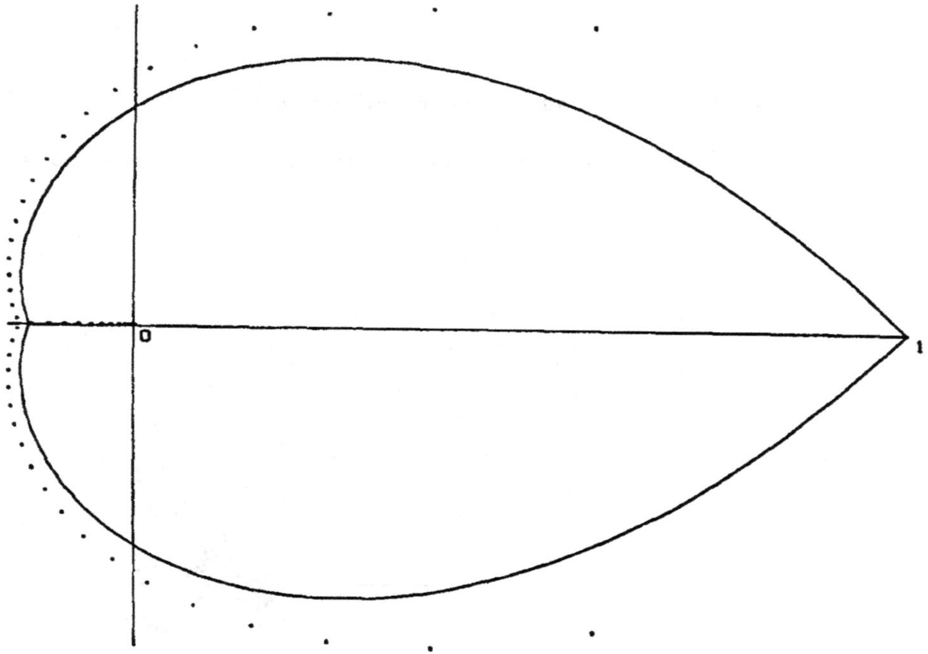

Figure 11. The Zeros of the Normalized Partial Sum $s_{50}(R_{50}z)$ for the Mittag-Leffler Function $E_{1/\lambda}(z)$ for $\lambda = 1/2$.

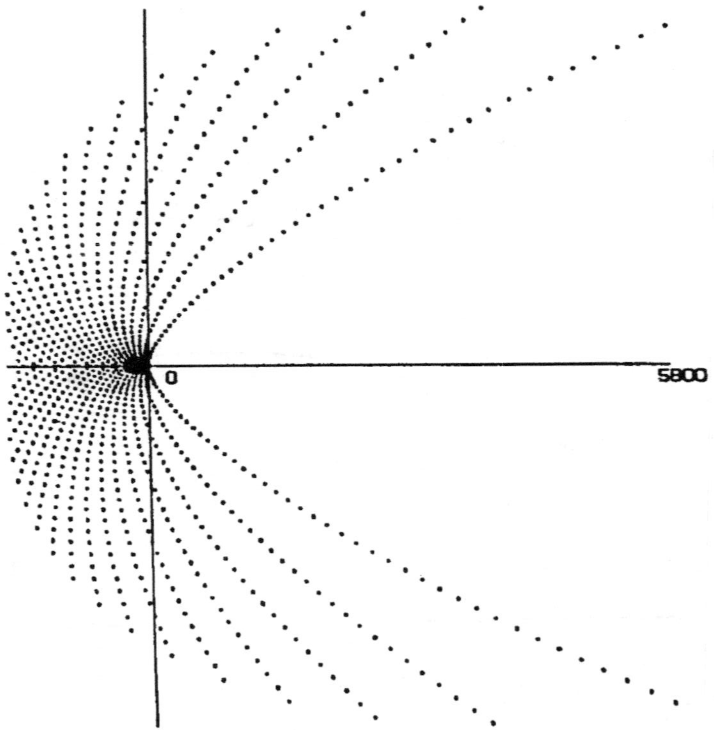

Figure 12. The Zeros of the Partial Sums $\{s_m(z)\}_{m=1}^{50}$ for the Mittag-Leffler Function $E_{1/\lambda}(z)$ for $\lambda = 1/2$.

4. Outline of the method. The method is quite general and may be applied to any transcendental entire function with positive Taylor coefficients. Rewrite (1.1) as

$$(4.1)\quad f(Rw) = \sum_{j=0}^{m} a_{m-j} R^{m-j} w^{m-j} + \sum_{j=1}^{\infty} a_{m+j} R^{m+j} w^{m+j},$$

where $R = R_m > 0$ and w is a complex variable.

The choice of a suitable relation between the suffix m and the associated R_m is an important step in the proofs.

Throughout this work we take

$$(4.2)\quad m = m(R)$$

to be the central index of the series (1.1), that is such that, for all nonnegative integers k,

$$(4.3)\quad a_m R^m \geq a_k R^k .$$

The radius R is not entirely determined by (4.3); in the case of the Mittag-Leffler function, every integer $m > 0$ is the central index associated with any R in the interval

$$(4.4)\quad \frac{a_{m-1}}{a_m} \leq R < \frac{a_m}{a_{m+1}} \qquad (a_j = 1/\Gamma(1+\tfrac{j}{\lambda}),\ 1 < \lambda < \infty) .$$

This follows from the fact that the sequence $\{a_{j-1}/a_j\}_{j=1}^{\infty}$ is strictly increasing (this will be established in (6.5) of §6).

In sections 4-17, we find it convenient to take

$$(4.5)\quad \log R_m = \frac{1}{\lambda} \log \left(\frac{m}{\lambda}\right) + \frac{1}{2m} \qquad (R = R_m),$$

as indicated in (2.2). In section 6, we verify that this choice does, in fact, satisfy the inequalities (4.4), for m large enough.

Introduce

$$(4.6)\qquad U_m(w) = \frac{f(Rw)}{a_m R^m w^m} \qquad (w \neq 0),$$

$$(4.7)\qquad Q_m(w) = \frac{s_m(Rw)}{a_m R^m w^m} = \sum_{j=0}^{m} b_{-j}(m)\, w^{-j},$$

$$(4.8)\qquad G_m(w) = \sum_{j=1}^{\infty} \frac{a_{m+j}}{a_m} R^j w^j = \sum_{j=1}^{\infty} b_j(m) w^j ,$$

where

$$(4.9)\qquad b_j(m) = \frac{a_{m+j}}{a_m} R^j \qquad (j \geq -m).$$

Since m is the central index, and since the a_j given in (1.3) are positive, we have

$$(4.10)\qquad 0 < b_j(m) \leq 1 \qquad (m \geq 1,\ j \geq -m);$$

we also note that by the properties of the Γ-function and (4.5),

$$(4.11)\qquad b_j(m) \to 1 \qquad (m \to +\infty,\ j = \text{fixed integer}).$$

From (4.10) and (4.11), it follows that

$$(4.12)\qquad G_m(w) \to \frac{w}{1-w} \qquad (m \to +\infty),$$

uniformly for all w in $|w| \leq 1 - \varepsilon$ $(0 < \varepsilon < 1)$. [A relation considerably more precise than (4.12) will be obtained in §9].

With the notations, (4.6), (4.7), and (4.8), we may rewrite (4.1) as

$$(4.13)\qquad Q_m(w) = U_m(w) - G_m(w) .$$

In the above relation, we take m large and fixed and deduce the relevant properties of $Q_m(w)$ from the following facts:

(i) by (4.12), $G_m(w)$ is approximately $w/(1-w)$, for $|w| < 1$;

(ii) the behavior of $U_m(w)$ may be studied with great precision by combining the well-known (and very accurate) asymptotic representation of the Mittag-Leffler function with the evaluation of $a_m R^m$ by Stirling's formula;

(iii) as a straightforward consequence of the Eneström-Kakeya theorem, $Q_m(w)$ has no zeros in the annulus $|w| \geq 1$.

The zeros of $Q_m(w)$ will be located by the method just described. In view of (4.7), our conclusions may be expressed in terms of the zeros of the sections $s_m(z)$; we are thus led to the Theorems 1-6 of this paper.

5. Notational conventions. We make little use of the symbols o, O, because they often mask some uniformity properties essential in our work.

Most of our error terms involve the factor ω which may be real or complex, but always satisfies the condition

$$|\omega| \leq 1.$$

The symbol ω may be a function of all the variables and parameters of the problem under consideration.

The symbols $m > m_0$, $R > R_0$, $k > k_0, \ldots$, following some formula, mean that the formula in question is only valid for large enough values of $m, R, k, \ldots$. The values of $m_0, R_0, k_0, \ldots$ are not the same ones at each occurrence.

The use of m_0 will be limited by the following additional restriction: the dependence on λ will not be indicated. Symbols, such as $m_0(h)$, $m_0(\varepsilon)$ indicate, respectively, a dependence on h and

λ, and on ε and λ. We remind the reader that λ (the order of the entire function under consideration) remains fixed throughout our proofs.

The symbols K and A denote, as usual, a positive constant and a positive absolute constant, respectively. If it is important, the parameters on which K may depend, will be explicitly mentioned. If clarity requires it, the various constants K and A which appear in our arguments will be fixed by the addition a subscript: $A_1, A_2, \ldots, K_1, K_2, \ldots$.

Finally, η_j is a member (real or complex) of a sequence and $\eta(r)$ is a function such that $\eta_j \to 0$ as $j \to \infty$, or $\eta(r) \to 0$ as $r \to \infty$. The use of η will be restricted to those cases where the limits, $\eta_j \to 0$ or $\eta(r) \to 0$, hold uniformly in some region explicitly mentioned by the context.

In spite of our efforts, the length of this manuscript has made it impossible to have the same symbol preserve its meaning throughout the paper. For example, Ω is used in two different senses in §14 and in §21. The same is true of E in §8 and in §21. We apologize to the reader for this inconvenience.

6. Properties of the Mittag-Leffler function of order $1 < \lambda < \infty$.

Here and in all sections through §17, we always write

$$(6.1)\qquad f(z) = E_{1/\lambda}(z) = \sum_{j=0}^{\infty} a_j z^j \;, \quad (a_j = 1/\Gamma(1+ \tfrac{j}{\lambda}));$$

the order λ $(1 < \lambda < +\infty)$ remains fixed throughout the paper.

We systematically use the following well-known asymptotic relations [5, p. 50]. If

$$|\arg z| \le \frac{3\pi}{4\lambda}, \qquad |z| \ge 2,$$

then

$$(6.2) \qquad f(z) = \lambda \exp(z^{\lambda}) - \frac{1}{z\Gamma\left(1 - \frac{1}{\lambda}\right)} + \omega \frac{\lambda^2 A_1}{|z|^2} \qquad (|\omega| \le 1).$$

If

$$\frac{3\pi}{4\lambda} \le |\arg z| \le 2\pi - \frac{3\pi}{4\lambda}, \qquad |z| \ge 2,$$

then

$$(6.3) \qquad f(z) = -\frac{1}{z\Gamma\left(1 - \frac{1}{\lambda}\right)} + \omega \frac{\lambda^2 A_2}{|z|^2} \qquad (|\omega| \le 1).$$

It is clear, by (6.1), that the properties of the Γ-function, and in particular Stirling's formula, will play a major role.

For our purposes, it suffices to remind the reader that

$$(6.4) \qquad \frac{d^2 \log\Gamma(x)}{dx^2} = \sum_{n=0}^{\infty} \frac{1}{(x+n)^2} > 0 \qquad (x > 0).$$

Thus, by the elements of calculus, (6.4) implies that

$$\Delta_2(x) = \log \Gamma\left(1 + \left(\frac{x+1}{\lambda}\right)\right) - 2 \log \Gamma\left(1 + \frac{x}{\lambda}\right) + \log \Gamma\left(1 + \frac{x-1}{\lambda}\right) > 0 \qquad (x > 0),$$

which directly yields that the sequence

$$(6.5) \qquad \left\{\frac{a_{j-1}}{a_j}\right\}_{j=1}^{\infty}$$

is strictly increasing.

We apply Stirling's expansion in the form

$$(6.6)\quad \log\Gamma(1+x) = \left(x+\frac{1}{2}\right)\log x - x + \frac{1}{2}\log 2\pi + \frac{1}{12x} + O\left(\frac{1}{x^3}\right) \qquad (x \to +\infty),$$

and deduce from it the approximations

$$(6.7)\quad \log\left(\frac{a_{m-1}}{a_m}\right) = \log\Gamma\left(1+\frac{m}{\lambda}\right) - \log\Gamma\left(1+\frac{m-1}{\lambda}\right) = \frac{1}{\lambda}\log\left(\frac{m}{\lambda}\right) + \frac{1}{2m}\left(1-\frac{1}{\lambda}\right) + O\left(\frac{1}{m^2}\right) \qquad (m \to +\infty),$$

and hence

$$(6.8)\quad \log\left(\frac{a_m}{a_{m+1}}\right) = \frac{1}{\lambda}\log\left(\frac{m}{\lambda}\right) + \frac{1}{2m}\left(1+\frac{1}{\lambda}\right) + O\left(\frac{1}{m^2}\right) \qquad (m \to +\infty).$$

A comparison of (6.7), (6.8), and (4.5) shows that

$$(6.9)\quad \frac{a_{m-1}}{a_m} < R_m < \frac{a_m}{a_{m+1}} \qquad (m > m_0).$$

We have thus verified that

$$m = m(R) = m(R_m)$$

is the central index corresponding to the value $R = R_m$. For the maximum term, we derive from (4.5) and (6.6) that

$$(6.10)\quad \log(a_m R^m) = \frac{m}{\lambda} - \frac{1}{2}\log\left(\frac{m}{\lambda}\right) + \frac{1}{2} - \frac{1}{2}\log 2\pi + \frac{\omega K}{m}$$
$$= R^\lambda - \frac{1}{2}\log(2\pi m/\lambda) + \frac{\omega K}{m} \qquad (m > m_0).$$

We also note that (6.9), (4.9), and the fact that the sequence a_{j-1}/a_j in (6.5) increases with j, implies

$$(6.11)\quad b_j(m) > b_{j+1}(m) \qquad (j = 0,1,2,\dots), \quad (m > m_0),$$

and

$$(6.12)\quad b_{-j}(m) > b_{-j-1}(m) \qquad (j = 0,1,\dots, m-1) \quad (m > m_0).$$

Stirling's formula (6.6) yields an approximation of $b_j(m)$: for any

(6.13) $\quad j \geq -\frac{m}{2}, \quad m > m_0,$

we have

(6.14) $\quad \log b_j(m) = -\delta(j) + \frac{\omega\lambda}{11m} \qquad (|\omega| \leq 1),$

where $\delta(x)$ is defined by

(6.15) $\quad \delta(x) = \frac{m+x}{\lambda} \log\left(1 + \frac{x}{m}\right) + \frac{1}{2} \log\left(1 + \frac{x}{m}\right) - x\left(\frac{1}{\lambda} + \frac{1}{2m}\right),$

$(x \in (-m, +\infty)).$

After some obvious reductions, we find

(6.16) $\quad \delta(x) = \frac{1}{\lambda}\int_0^x \log\left(1 + \frac{t}{m}\right) dt + \frac{1}{2}\left(\log\left(1 + \frac{x}{m}\right) - \frac{x}{m}\right) \qquad (x \geq 0).$

In view of the elementary approximation

(6.17) $\quad \log(1 + h) = h - \frac{h^2}{2} + \omega h^3 \qquad (|h| \leq \frac{1}{2}),$

(6.16) yields

(6.18) $\quad \delta(x) = \frac{x^2}{2\lambda m} - \frac{x^3}{6\lambda m^2} + \frac{\omega x^4}{4\lambda m^3} + \frac{\omega x^2}{2m^2} \qquad (0 \leq x \leq \frac{m}{2}).$

Introduce the auxiliary integer

(6.19) $\quad L = \left[3(\lambda m \log m)^{1/2}\right].$

This choice of L (as well as the choice of the factor 21 which appears in the equation below) is unmotivated at this stage, and will be justified in §7. However, from a straightforward computation using (6.18) and (6.19), we deduce

(6.20) $\quad \delta(L+1) - \frac{L+1}{21}\left(\frac{\log m}{\lambda m}\right)^{1/2} > 4 \log m \qquad (m > m_0).$

For

(6.21) $\quad x \geq L + 1,$

equation (6.16) yields

$$\delta(x) - \delta(L+1) \geq \frac{1}{\lambda}(x - L - 1)\log\left(1 + \frac{L+1}{m}\right) - \frac{1}{2m}(x - L - 1)$$

$$\geq \frac{5}{2}(x - L - 1)\left(\frac{\log m}{\lambda m}\right)^{1/2} \qquad (m > m_0) \quad ,$$

and

$$(6.22) \qquad \delta(x) - \delta(L+1) - \left(\frac{x-L-1}{21}\right)\left(\frac{\log m}{\lambda m}\right)^{1/2} \geq 2(x-L-1)\left(\frac{\log m}{\lambda m}\right)^{1/2} \qquad (m > m_0).$$

Replace in (6.15) x by -x; this leads to

$$(6.23) \qquad \delta_1(x) = \delta(-x) = \frac{m-x}{\lambda}\log\left(1 - \frac{x}{m}\right) + \frac{1}{2}\log\left(1 - \frac{x}{m}\right) + x\left(\frac{1}{\lambda} + \frac{1}{2m}\right) \qquad (0 \leq x < m).$$

The analogue of (6.16) is

$$(6.24) \qquad \delta_1(x) = -\frac{1}{\lambda}\int_0^x \log\left(1 - \frac{t}{m}\right)dt + \frac{1}{2}\left(\log\left(1 - \frac{x}{m}\right) + \frac{x}{m}\right) \qquad (0 \leq x < m),$$

and of (6.18),

$$(6.25) \qquad \delta_1(x) = \frac{x^2}{2\lambda m} + \frac{x^3}{6\lambda m^2} + \frac{\omega x^4}{4\lambda m^3} + \frac{\omega x^2}{2m^2} \qquad (0 \leq x \leq \tfrac{m}{2}).$$

As in (6.20), we similarly have

$$(6.26) \qquad \delta_1(L+1) - \frac{(L+1)}{21}\left(\frac{\log m}{\lambda m}\right)^{1/2} > 4\log m \qquad (m > m_0).$$

To obtain the analogue of (6.22) we notice that (6.24) implies

$$\delta_1(x) - \delta_1(L+1) \geq \frac{x - L - 1}{\lambda}\log\left(\frac{1}{1 - \frac{L+1}{m}}\right) + \frac{1}{2}\log\left(\frac{m - x}{m - L - 1}\right)$$

$$\geq \frac{(x - L - 1)\ 3(\lambda m)^{1/2}(\log m)^{1/2}}{\lambda m} - A\frac{x - L - 1}{m} \geq \frac{5}{2}\frac{(x - L - 1)(\log m)^{1/2}}{(\lambda m)^{1/2}}$$

$$(L + 1 \leq x \leq \tfrac{m}{2}\ ,\ m > m_0)$$

and hence

$$(6.27)\qquad \delta_1(x) - \delta_1(L+1) - \left(\frac{x-(L+1)}{21}\right)\left(\frac{\log m}{\lambda m}\right)^{1/2}$$
$$\geq 2(x-L-1)\left(\frac{\log m}{\lambda m}\right)^{1/2} \quad (L+1 \leq x \leq \tfrac{m}{2}\,,\ m > m_0).$$

We shall also make use of

$$(6.28)\qquad \delta_1\left(\left[\frac{m}{2}\right]\right) > \frac{m}{8\lambda} \qquad (m > m_0).$$

7. Estimates for $G_m(w)$ and $Q_m(w)$. We propose to study the behavior of both functions on the circumference $|w| = 1$, for large values of m, where $G_m(w)$ and $Q_m(w)$ are defined, respectively, in (4.8) and (4.7).

Since the immediate vicinity of the point $w = 1$ is of particular interest, we estimate $G_m(w)$ for

$$(7.1)\qquad |w| \leq \exp\left\{\frac{1}{21}\left(\frac{\log m}{\lambda m}\right)^{1/2}\right\} \qquad (m > m_0),$$

and $Q_m(w)$ for

$$(7.2)\qquad |w| \geq \exp\left\{-\frac{1}{21}\left(\frac{\log m}{\lambda m}\right)^{1/2}\right\} \qquad (m > m_0).$$

The constant 21 which appears in (7.1) and (7.2) has been selected for simplicity. Other choices would serve our purposes equally well. Defining L by (6.19), and restricting w by (7.1), we deduce from (4.8) and (6.14) that

$$(7.3)\qquad G_m(w) = \sum_{j=1}^{L} \exp\left(-\delta(j) + \frac{\omega\lambda}{11m}\right) w^j + \sum\nolimits_2 \ ,$$

where

$$(7.4)\quad \left|\sum_2\right| \le \exp\left(\frac{\lambda}{11m}\right) \sum_{j=L+1}^{\infty} \exp\left(-\delta(j) + \frac{j}{21}\left(\frac{\log m}{\lambda m}\right)^{1/2}\right).$$

By (6.20) and (6.22), we find

$$\left|\sum_2\right| \le 2m^{-4} \sum_{k=0}^{\infty} \exp\left(-2k\left(\frac{\log m}{\lambda m}\right)^{1/2}\right) < m^{\frac{1}{2}-4} = m^{-7/2} \qquad (m > m_0).$$

Let Σ_1 denote the first sum in the right-hand side of (7.3); in view of (6.18), (6.19), and (7.1), we find

$$(7.5)\quad \left|\sum_1 - \sum_{j=1}^{L} \exp\left(-\frac{j^2}{2\lambda m} + \frac{j^3}{6\lambda m^2}\right) w^j\right|$$

$$\le \frac{21\lambda(\log m)^2}{m} \sum_{j=1}^{L} \exp\left(-\frac{j^2}{2\lambda m} + \frac{j^3}{6\lambda m^2}\right) |w|^j$$

$$\le 21\lambda\, m^{-1}(\log m)^2 \sum_{j=1}^{L} \exp\left(-\frac{j^2}{2\lambda m}\right) \exp\left(\frac{L^3}{6\lambda m^2} + \frac{L(\log m)^{1/2}}{21(\lambda m)^{1/2}}\right)$$

$$\le 22\lambda\, m^{-1+\frac{1}{7}}(\log m)^2 \sum_{j=1}^{L} \exp\left(-\frac{j^2}{2\lambda m}\right) \qquad (m > m_0).$$

For the last sum in (7.5), we use the elementary estimate

$$(7.6)\quad \sum_{j=1}^{\infty} \exp\left(-\frac{j^2}{2\lambda m}\right) < \int_0^{\infty} \exp\left(-\frac{t^2}{2\lambda m}\right) dt = (\pi\lambda m/2)^{1/2},$$

and note, for later use, the generalization

$$(7.7)\quad \sum_{j=1}^{\infty} j^{\alpha} \exp\left(-\frac{j^2}{2\lambda m}\right) < 1 + \int_1^{\infty} (t+1)^{\alpha} \exp\left(-\frac{t^2}{2\lambda m}\right) dt$$

$$< 1 + 2^{\alpha}(2\lambda m)^{\frac{\alpha+1}{2}} \int_0^{\infty} t^{\alpha} \exp(-t^2)\,dt \qquad (\alpha > 0).$$

Hence (7.5) takes the simpler form

(7.8) $$\left|\sum_1 - \sum_{j=1}^{L} \exp\left(-\frac{j^2}{2\lambda m} + \frac{j^3}{6\lambda m^2}\right) w^j\right| \le \frac{1}{2} m^{-1/3} \qquad (m > m_0).$$

The final step of our evaluation is obvious:

(7.9) $$\sum_{j=1}^{L} \exp\left(-\frac{j^2}{2\lambda m} + \frac{j^3}{6\lambda m^2}\right) w^j = \sum_{j=1}^{L} \exp\left(-\frac{j^2}{2\lambda m}\right) w^j$$

$$+ \frac{1}{6\lambda m^2} \sum_{j=1}^{L} j^3 \exp\left(-\frac{j^2}{2\lambda m}\right) w^j$$

$$+ \frac{K\omega}{m^4} \exp\left(\frac{L}{21}\left(\frac{\log m}{\lambda m}\right)^{1/2}\right) \sum_{j=1}^{L} j^6 \exp\left(-\frac{j^2}{2\lambda m}\right).$$

By (7.7) and (6.19), the error term in (7.9) may be written as

(7.10) $$K\omega m^{-\frac{1}{2}+\frac{1}{7}} \qquad (m > m_0).$$

Combining the information contained in (7.3) – (7.10), we finally obtain

(7.11) $$G_m(w) = \sum_{j=1}^{L} \exp\left(-\frac{j^2}{2\lambda m}\right) w^j$$

$$+ \frac{1}{6\lambda m^2} \sum_{j=1}^{L} j^3 \exp\left(-\frac{j^2}{2\lambda m}\right) w^j + \omega m^{-1/3} \qquad (m > m_0).$$

This formula holds uniformly for all w restricted by (7.1). (The quantity L is defined in (6.19).)

The treatment of $Q_m(w)$ $\left(\text{with } |w| \ge \exp\left(-\frac{(\log m)^{1/2}}{21(\lambda m)^{1/2}}\right)\right)$ follows the same lines as the preceding one of $G_m(w)$. We start from (4.7) and (6.14) and on writing $k(m) = [m/2]$, we find

$$(7.12)\qquad Q_m(w) = \sum_{j=0}^{k(m)-1} \exp\left(-\delta(-j) + \frac{\omega\lambda}{11m}\right) w^{-j} + \sum_3 ,$$

where

$$\sum_3 = \sum_{j=k(m)}^{m} b_{-j} w^{-j} .$$

By (6.12) and (7.2),

$$\left|\sum_3\right| < m b_{-k(m)} \exp\left(\frac{m}{21}\left(\frac{\log m}{\lambda m}\right)^{1/2}\right) ,$$

and taking (6.28) into account, we find

$$(7.13)\qquad \left|\sum_3\right| < 2m \exp\left(-\frac{m}{8\lambda} + \frac{1}{21}\left(\frac{m \log m}{\lambda}\right)^{1/2}\right) < \frac{1}{m} \qquad (m > m_0).$$

With the definition of $\delta_1(x)$ (in (6.23)), (7.12) and (7.13) show that the analogue of (7.3) is now

$$(7.14)\qquad Q_m(w) = \sum_{j=0}^{L} \exp\left(-\delta_1(j) + \frac{\omega\lambda}{11m}\right) w^{-j} + \sum_3 + \sum_4$$

with

$$\left|\sum_4\right| \le \exp\left(\frac{\lambda}{11m}\right) \sum_{j=L+1}^{k(m)-1} \exp\left(-\delta_1(j) + \frac{j}{21}\left(\frac{\log m}{\lambda m}\right)^{1/2}\right) .$$

From this point on, we use the relations (6.24) - (6.27) instead of the corresponding relations for $\delta(x)$. We thus find the analogue of (7.11) to be

$$(7.15)\qquad Q_m(w) = \sum_{j=0}^{L} \exp\left(-\frac{j^2}{2\lambda m}\right) w^{-j} - \frac{1}{6\lambda m^2} \sum_{j=0}^{L} j^3 \exp\left(-\frac{j^2}{2\lambda m}\right) w^{-j}$$
$$+ \omega m^{-1/3} \qquad \left(m > m_0,\ L = \left[3(\lambda m \log m)^{1/2}\right]\right),$$

valid uniformly for all w satisfying (7.2).

8. A differential equation. With L as defined in (6.19), we introduce

$$(8.1)\qquad J_m(w) = \sum_{j=1}^{L} e_j w^j \qquad \left(e_j = \exp\left(-\frac{j^2}{2\lambda m}\right)\right).$$

Assume for simplicity that

$$(8.2)\qquad |w| \le \exp(B(2/\lambda m)^{1/2}) \qquad (1 < B),$$

where the constant B may be chosen as large as we wish. This condition, which could be relaxed, suffices for our purposes. Obviously,

$$(8.3)\qquad (1-w)J_m(w) - w = w(e_1-1) + \sum_{j=2}^{L} (e_j-e_{j-1})w^j - e_L w^{L+1}.$$

Now

$$(8.4)\qquad e_j-e_{j-1} = e_j\left(1 - \exp\left(\frac{2j-1}{2\lambda m}\right)\right) = -e_j\left\{\frac{j}{\lambda m} - \frac{1}{2\lambda m} + f_j\right\},$$

with

$$(8.5)\qquad f_j = \sum_{k=2}^{\infty} \frac{(2j-1)^k}{k!(2\lambda m)^k} = \omega \frac{j^2}{(\lambda m)^2} \qquad (1 \le j \le L,\ m > m_0).$$

Using (8.4) and (8.5) in (8.3), we find

$$(8.6)\qquad w-(1-w)J_m(w) = \frac{w}{\lambda m}\sum_{j=1}^{L} j e_j w^{j-1} + wE_m(w) = \frac{wJ_m'(w)}{\lambda m} + wE_m(w),$$

where $E_m(w)$ is a polynomial given by

$$(8.7)\qquad E_m(w) = (1-e_1-(e_1/\lambda m)) + e_L w^L - \frac{1}{2\lambda m}\sum_{j=2}^{L} e_j w^{j-1} + \sum_{j=2}^{L} e_j f_j w^{j-1}.$$

Hence

$$(8.8) \qquad |E_m(w)| < \frac{1}{\lambda m} + \left(\frac{1}{2\lambda m}\sum_{j=2}^{\infty} e_j + \frac{1}{\lambda^2 m^2}\sum_{j=2}^{\infty} j^2 e_j + e_L\right) \cdot \max(1, |w|^L),$$

with

$$(8.9) \qquad 0 < e_L < e^3\, m^{-9/2} \qquad (m \geq 3),$$

and, in view of (8.2),

$$(8.10) \qquad |w|^L < \exp(5B(\log m)^{1/2}) = \tilde{\eta}(m) \qquad (m > m_0).$$

It is important to note that, for any $\varepsilon > 0$,

$$(8.11) \qquad \tilde{\eta}(m)m^{-\varepsilon} \to 0 \qquad (m \to +\infty),$$

and hence

$$(8.12) \qquad e_L|w|^L < m^{-4} < \frac{1}{\lambda m} \qquad (m > m_0).$$

Using (7.6), (7.7), (8.10), and (8.12) in (8.8), we find

$$(8.13) \qquad |E_m(w)| < A(\lambda m)^{-\frac{1}{2}} \qquad (|w| \leq 1,\ m > m_0),$$

and, under the less restrictive assumption (8.2)

$$(8.14) \qquad |E_m(w)| < A(\lambda m)^{-\frac{1}{2}}\,\tilde{\eta}(m) \qquad (m > m_0).$$

Consider $E_m(w)$ as a known function and treat (8.6) as a differential equation defining $J_m(w)$; this leads to

$$(8.15) \qquad J_m' + \lambda m\left(\frac{1}{w} - 1\right) J_m = \lambda m(1-E_m).$$

The integration of (8.15) is elementary and explicit: using the integrating factor $w^{\lambda m}e^{-\lambda m w}$, we find that

$$(8.16)\qquad J_m(w) = \lambda m\ \exp(\lambda m w)w^{-\lambda m}\int_1^w \exp(-\lambda m t)t^{\lambda m}(1-E_m(t))dt$$

$$+ J_m(1)\ \exp(\lambda m(w-1))w^{-\lambda m}.$$

The quantity $J_m(1)$ needs to be evaluated with some precision. By definition (cf.(8.1))

$$(8.17)\qquad J_m(1) = \sum_{j=1}^{L} e_j = \int_1^{L+1} \exp\left(-\frac{t^2}{2\lambda m}\right) dt$$

$$+\sum_{j=1}^{L}\int_j^{j+1}\left\{\exp\left(-\frac{j^2}{2\lambda m}\right) - \exp\left(-\frac{t^2}{2\lambda m}\right)\right\} dt$$

$$= \int_0^{+\infty} \exp\left(-\frac{t^2}{2\lambda m}\right) dt$$

$$-\left(\int_0^1 + \int_{L+1}^{+\infty}\right) \exp\left(-\frac{t^2}{2\lambda m}\right) dt + \omega\left\{\exp\left(-\frac{1}{2\lambda m}\right)\right.$$

$$\left. - \exp\left(-\frac{(L+1)^2}{2\lambda m}\right)\right\} = \left(\frac{\pi\lambda m}{2}\right)^{1/2} + 3\omega \qquad (m > m_0).$$

Combining (8.16) and (8.17), we find

$$(8.18)\qquad J_m(w) = \exp(\lambda m(w-1))w^{-\lambda m}\left\{\lambda m\int_1^w \exp(-\lambda m(t-1))t^{\lambda m}(1-E_m(t))dt\right.$$

$$\left. + \left(\frac{\pi\lambda m}{2}\right)^{1/2} + 3\omega\right\} \qquad (|w| \le \exp(B(2/\lambda m)^{1/2}),\ m > m_0).$$

We now proceed to deduce from this explicit form of $J_m(w)$ a good approximation, valid for small values of w-1. More precisely, we take

$$(8.19) \qquad w-1 = \rho\left(\frac{2}{\lambda m}\right)^{1/2} e^{i\theta} \qquad (0 \le \rho \le B,\ \theta \text{ real}).$$

Throughout the remainder of the section, we always take

$$(8.20) \qquad m > \tilde{m}_0 > 0,$$

and $\tilde{m}_0$ large enough to imply

$$(8.21) \qquad B(2/\lambda\tilde{m}_0)^{1/2} < \frac{1}{2}.$$

We need the following consequence of (8.19) and (8.20):

$$(8.22) \qquad \exp(\lambda m(w-1))w^{-\lambda m} = \exp\left(\lambda m \sum_{k=2}^{\infty} \frac{(1-w)^k}{k}\right)$$

$$= \exp\left(\rho^2 e^{2i\theta} + \frac{\omega 2B^3}{(\lambda m)^{1/2}}\right) \qquad (m > \tilde{m}_0).$$

Perform in (8.18) the change of variable

$$(8.23) \qquad t-1 = e^{i\theta}(2/\lambda m)^{1/2} s.$$

Taking into account (8.14) and (8.22), we deduce from (8.18) (after some obvious reductions)

$$(8.24) \qquad J_m(1 + \rho(2/\lambda m)^{1/2} e^{i\theta})$$

$$= (2\lambda m)^{1/2} \exp\left(\rho^2 e^{2i\theta} + \frac{\omega 2B^3}{(\lambda m)^{1/2}}\right) \left\{\frac{\pi^{1/2}}{2} + 3\omega(\lambda m)^{-1/2}\right.$$

$$\left. + e^{i\theta}\int_0^{\rho} \exp\left(-s^2 e^{2i\theta} + \frac{\omega 2B^3}{(\lambda m)^{1/2}}\right)\left(1 + \frac{\omega A\tilde{\eta}(m)}{(\lambda m)^{1/2}}\right) ds\right\}$$

$$(m > \tilde{m}_0 + m_0).$$

To analyze the error terms in (8.24), we systematically use the elementary inequality

(8.25) $$|e^{\zeta}-1| \leq |\zeta| e^{|\zeta|}.$$

In view of our notational conventions, we readily see that (8.24) may be given the more transparent form

(8.26) $$J_m(1 + \rho(2/\lambda m)^{1/2} e^{i\theta})$$

$$= (2\lambda m)^{1/2} \exp(\rho^2 e^{2i\theta}) \left\{ \frac{\pi^{1/2}}{2} + e^{i\theta} \int_0^{\rho} \exp(-s^2 e^{2i\theta}) ds \right.$$

$$\left. + \omega K(\lambda, B) m^{-1/2} \tilde{\eta}(m) \right\} \qquad (0 \leq \rho \leq B,\ m > m_0(B)).$$

As an immediate consequence of (8.26), we obtain our fundamental

Lemma 8.1. As $m \to \infty$, we have, uniformly on every compact subset of the ζ-plane ($\zeta = \rho e^{i\theta}$),

(8.27) $$\left(\frac{2}{\pi\lambda m}\right)^{1/2} J_m\left(1 + \left(\frac{2}{\lambda m}\right)^{1/2} \zeta\right) \to e^{\zeta^2}\left(1 + \frac{2}{\sqrt{\pi}} \int_0^{\zeta} e^{-t^2}\, dt\right)$$

$$= e^{\zeta^2} \operatorname{erfc}(-\zeta),$$

as well as

(8.28) $$\left(\frac{2}{\pi\lambda m}\right)^{1/2} Q_m\left(\frac{1}{w}\right) \to e^{\zeta^2}\left(1 + \frac{2}{\sqrt{\pi}} \int_0^{\zeta} e^{-t^2}\, dt\right) = e^{\zeta^2} \operatorname{erfc}(-\zeta),$$

where $w = 1 + (2/\lambda m)^{1/2} \zeta$.

Proof. With the definition of $\operatorname{erfc}(\zeta)$ in (2.12), the relations (8.26), (8.10), and (8.11) clearly yield (8.27), and do in fact give us a simple uniform bound for the difference between the two sides of (8.27). Relation (8.28) follows similarly from (8.27), (8.11), (8.10), (7.7) and (7.15). This completes the proof of the lemma.

The classical error function

$$\frac{2}{\sqrt{\pi}}\int_0^{\zeta} e^{-t^2}\,dt = H(\zeta),$$

which appears in (8.27) and (8.28), is an odd entire function of order two. From the point of view of Nevanlinna theory, it has precisely two deficient values 1 and -1. Both deficiences are equal to 1/2, so that their sum is extremal in Nevanlinna's fundamental deficiency relation. We also note that 1 and -1 are asymptotic values:

$$\lim_{x \to +\infty} H(x) = - \lim_{x \to +\infty} H(-x) = 1$$

Following Nevanlinna, we define (cf. (8.28))

(8.29) $$n\left(t, \frac{1}{H(\zeta)+1}\right) = n\left(t, \frac{1}{\operatorname{erfc}(-\zeta)}\right) = \nu(t)$$

to be the number of zeros of $1 + H(\zeta) = \operatorname{erfc}(-\zeta)$ in the disk $|\zeta| \leq t$, and we set

(8.30) $$N\left(\rho, \frac{1}{\operatorname{erfc}(-\zeta)}\right) \equiv \int_0^{\rho} \frac{\nu(t)}{t}\,dt.$$

Nevanlinna proves [20, pp. 19-21] that

(8.31) $$\int_0^{\rho} \frac{\nu(t)}{t}\,dt \sim \frac{\rho^2}{2\pi} \qquad (\rho \to +\infty)$$

and hence, by a straightforward tauberian argument (cf. [35, p. 40] for a similar case),

(8.32) $$\nu(\rho) \sim \frac{\rho^2}{\pi} \qquad (\rho \to +\infty).$$

With regard to the distribution of roots of the equation

(8.33) $$\operatorname{erfc}(-\zeta) = H(\zeta) + 1 = 0,$$

we mention two elementary facts.

I. All the roots of the equation (8.33) lie in the half plane $\operatorname{Re} \zeta > 0$.

II. If

$$H(\zeta_j) + 1 = 0 \qquad (\zeta_j = \rho_j e^{i\theta_j})$$

and if

$$0 \leq \theta_j < \frac{\pi}{2},$$

then

$$\theta_j \to \frac{\pi}{4}$$

as $\rho_j \to +\infty$.

It is unnecessary to prove these properties because they are not needed to establish Theorem 1. Nevertheless we find it interesting to point out that the property I is a simple consequence of (8.27). To see this, assume that

$$H(\zeta_1) + 1 = 0$$

for some $\zeta_1 = \rho_1 e^{i\theta_1}$ with

$$\text{(8.34)} \qquad \frac{\pi}{2} < \theta_1 < \frac{3\pi}{2}, \qquad \rho_1 > 0.$$

Then, by (8.27) and Hurwitz' theorem, there exists a sequence η_m such that $\eta_m \to 0$ $(m \to +\infty)$ and such that

$$J_m\left(1 + \left(\frac{2}{\lambda m}\right)^{1/2} (\zeta_1 + \eta_m)\right) = 0 \qquad (m > m_0).$$

If m is large enough, we have, by (8.34),

$$\left|1 + \left(\frac{2}{\lambda m}\right)^{1/2} (\zeta_1 + \eta_m)\right| < 1,$$

and there would exist polynomials in w of the form $J_m(w)/w$ which would have a zero inside the unit disk. This contradicts the Eneström-Kakeya theorem because the polynomial $J_m(w)/w$ has its coefficients positive and decreasing. From this contradiction, we conclude that the inequalities (8.34) cannot hold. That the equation (8.33) has no roots on the imaginary axis is obvious because

$$\mathrm{Re}\left\{H(iy)\right\} = 0 \qquad (-\infty < y < +\infty).$$

The proof of assertion I is now complete.

9. Estimates for $J_m(w)$ near the circumference $|w|=1$. If $(w-1)$ is not very small, an integration of the differential equation (8.15) becomes unnecessary. It is much simpler, in this case, to derive good approximations from the following

Lemma 9.1. Put

$$(9.1) \qquad J_m(w;\alpha) = \sum_{j=1}^{L} j^{\alpha}\exp\left(-\frac{j^2}{2\lambda m}\right)w^j \qquad (0<\alpha,\ 0<m).$$

Then, if $|w| \leq 1$, we have

$$(9.2) \qquad |(1-w)J_m(w;\alpha)| \leq 4(e^{-1}\alpha\lambda m)^{\alpha/2}.$$

Proof. Write

$$g_j = j^{\alpha}\exp\left(-\frac{j^2}{2\lambda m}\right) = g(j),$$

where, for simplicity, we have omitted from the notations g_j, $g(j)$, the parameters α and m.

Consider $g(j)$ as a continuous function of the real variable j; since $\alpha>0$, $g(j)$ increases in the interval

$$0 \leq j \leq (\alpha\lambda m)^{1/2}$$

and decreases for $j \geq (\alpha\lambda m)^{1/2}$. Hence

$$g(j) \leq (e^{-1}\alpha\lambda m)^{\alpha/2} \qquad (j \geq 0,\ \alpha > 0).$$

By definition

$$(9.3) \qquad (1-w)J_m(w;\alpha) = g_1 w + \sum_{j=1}^{L-1} (g_{j+1}-g_j)w^{j+1} - g_L w^{L+1},$$

and consequently

$$(9.4) \qquad |(1-w)J_m(w;\alpha)| \leq \{g_1 + \sum_{j=1}^{L-1} |g_{j+1}-g_j| + g_L\} \qquad (|w| \leq 1).$$

Consider the integer

$$s = [(\alpha\lambda m)^{1/2}]$$

and assume first that

$$(9.5) \qquad 1 < s < L-1.$$

It is then obvious that

$$(9.6) \quad g_1 + \sum_{j=1}^{L-1} |g_{j+1}-g_j| + g_L = g_s + |g_{s+1}-g_s| + g_{s+1} \leq 4(e^{-1}\alpha\lambda m)^{\alpha/2}.$$

Using (9.6) in (9.4) we obtain (9.2) under the restrictions (9.5). It is immediately seen that (9.2) is not affected if the latter restrictions are omitted. The proof of Lemma 9.1 is now complete.

Lemma 9.2. *Let* δ $(0 < \delta < 1)$ *be given. Then*

$$(9.7) \qquad G_m(w) = \frac{w}{1-w} + 2\omega m^{-1/3},$$

uniformly for w *and* m *restricted by*

(9.8) $$|w-1| \geq \delta , \qquad |w| \leq 1 , \qquad (m > m_o(\delta)) .$$

Proof. By (8.6), (8.13) and (9.2) (with $\alpha=1$), we find

$$J_m(w)(w-1) + w = \omega|1-w|^{-1} 4(e\lambda m)^{-1/2} + \omega A(\lambda m)^{-1/2} ,$$

and hence, by (9.8),

(9.9) $$J_m(w) = \frac{w}{1-w} + \omega A(\lambda m)^{-1/2} \delta^{-2} .$$

To complete the proof of (9.7) we combine (9.9), (7.11) and use the following simple consequence of (9.2) (with $\alpha = 3$):

$$\frac{|J_m(w;3)|}{6\lambda m^2} \leq A|1-w|^{-1} \lambda^{1/2} m^{-1/2} .$$

10. Existence and uniqueness of the Szegö curve.

Let

(10.1) $$\xi = |\xi|e^{i\phi} = \xi(\phi) \qquad (\xi \neq 0) ,$$

be a complex variable and consider, for $\phi \left(|\phi| \leq \frac{\pi}{2\lambda}\right)$ fixed, the expression

(10.2) $$X(|\xi|) = |\xi|^{\lambda}\cos(\phi\lambda) - 1 - \lambda \log|\xi| \qquad (0 < |\xi| \leq 1) .$$

As $|\xi|$ increases from 0 to 1, $X(|\xi|)$ decreases from $+\infty$ to $\cos(\phi\lambda) - 1 \leq 0$. The decreasing character of $X(|\xi|)$ is immediate since for ϕ fixed

$$\frac{dX}{d|\xi|} = \frac{\lambda}{|\xi|}\left(|\xi|^{\lambda}\cos(\phi\lambda) - 1\right) < 0 \qquad (0 < |\xi| < 1) .$$

It is now clear that there exists one and only one $|\xi|$ such that

(10.3) $$0 < |\xi| = |\xi(\phi)| \leq 1, \qquad X(|\xi|) = 0 , \quad |\phi| \leq \frac{\pi}{2\lambda} .$$

We have thus verified the existence and uniqueness of $\sigma(\phi)$ in part (i) of the definition of the normalized Szegö curve S given in §2.(I).

The circular portion of S:

$$\xi(\phi) = \exp\left(-\frac{1}{\lambda} + i\phi\right) \qquad \left(\frac{\pi}{2\lambda} \leq \phi \leq 2\pi - \frac{\pi}{2\lambda}\right)$$

requires no discussion.

11. Crude estimates for $|U_m(w)|$ and $|Q_m(w)|$. We first make use of the asymptotic relation (6.2); we set

(11.1) $$A^* \equiv \frac{1}{4}(\log 3 - 1) > 0 ,$$

then there is a constant $B_o = B_o(\lambda) > 1$ such that

(11.2) $$\log|f(z)| = \mathrm{Re}(z^{\lambda}) + \log\lambda + A^*\omega(z) \qquad (|\omega(z)| \leq 1) ,$$

uniformly for

(11.3) $$|\arg z| \leq \pi/2\lambda , \qquad |z| \geq B_o .$$

Let $\xi = \xi(\phi) = \sigma(\phi)e^{i\phi}$ be any point of the normalized Szegö curve and let t be a positive variable which satisfies

$$t > e^{1/\lambda}B_o/R ,$$

where $R = R_m$ is defined as in (2.2). Then we have

$$|Rt\xi(\phi)| > B_o , \tag{11.4}$$

and (11.2) is valid with

$$z = Rt\xi(\phi) = Rw , \quad |\phi| \leq \pi/2\lambda . \tag{11.5}$$

Hence, for $|\phi| \leq \pi/2\lambda$,

$$\log|f(Rt\xi(\phi))| = (Rt\sigma(\phi))^{\lambda}\cos(\phi\lambda) + \log\lambda + A^*\omega . \tag{11.6}$$

Also, by (2.2),

$$R^{\lambda} = \frac{m}{\lambda}\left\{1 + \frac{\lambda}{2m} + \frac{K\omega}{m^2}\right\} \quad (m \geq 1) . \tag{11.7}$$

Taking into account (4.6), (6.10) and (10.2), we deduce from (11.6) and (11.7) that for $m > m_o$,

$$\begin{aligned}\log|U_m(w)| &= \frac{m}{\lambda}\left\{(t\sigma)^{\lambda}\cos(\phi\lambda) - 1 - \lambda\log(t\sigma)\right\} \\ &\quad + \frac{1}{2}\left\{(t\sigma)^{\lambda}\cos(\phi\lambda) - 1\right\} + \frac{1}{2}\log(2\pi m\lambda) + 2A^*\omega ,\end{aligned} \tag{11.8}$$

that is,

$$\log|U_m(w)| = \frac{m}{\lambda}\, X(|w|) + \frac{1}{2}\left\{|w|^{\lambda}\cos(\phi\lambda) - 1\right\} + \frac{1}{2}\log(2\pi m\lambda) + 2A^*\omega . \tag{11.9}$$

From the definition of A* in (11.1) and the fact that $X(|w|) \geq X(\sigma) = 0$ for $|w| \leq \sigma$, equation (11.9) yields

$$\log|U_m(w)| \geq 0 - \frac{1}{2} + \frac{1}{2}\log(2\pi m\lambda) - \frac{1}{2}(\log 3 - 1)$$

or

$$|U_m(w)| \geq \left(\frac{2\pi m\lambda}{3}\right)^{1/2} \qquad (m > m_0) \tag{11.10}$$

provided $w \in \Sigma_1(m)$, where

$$\Sigma_1(m) = \left\{ w = |w|e^{i\phi} : B_0/R_m \leq |w| \leq \sigma(\phi), \quad |\phi| \leq \pi/2\lambda \right\} . \tag{11.11}$$

Next, let $h > 0$ be a given fixed quantity and suppose that

$$(1 + h)\sigma(\phi) \leq 1 . \tag{11.12}$$

Since $X(\sigma(\phi)) = 0$, we have from (10.2)

$$X(t\sigma(\phi)) = (1 + \lambda\log\sigma(\phi))(t^\lambda - 1) - \lambda\log t , \tag{11.13}$$

and hence, in view of (11.12),

$$X((1+h)\sigma(\phi)) \leq (1-\lambda\log(1+h))((1+h)^\lambda-1)-\lambda\log(1+h) . \tag{11.14}$$

Denote by $Y(h)$ the right-hand side of (11.14). Then

$$Y'(h) = -\lambda^2(1+h)^{\lambda-1}\log(1+h) ,$$

$$Y''(h) = -\lambda^2(1+h)^{\lambda-2}((\lambda-1)\log(1+h)+1) ,$$

and by Taylor's formula

$$Y(h) = \frac{Y''(h')}{2}h^2 \qquad (0 \leq h' \leq h < \tfrac{1}{2}) . \tag{11.15}$$

Since

$$Y''(h') \leq -\lambda^2(1+h')^{\lambda-2} \qquad (\lambda > 1) ,$$

we deduce from (11.14), (11.15) and the decreasing character of X

$$X(|w|) \leq X((1+h)\sigma(\phi)) \leq Y(h) \leq -\frac{\lambda^2}{2}h^2(1+h')^{-1} \leq -\frac{\lambda^2h^2}{3} \quad (0\leq h\leq\tfrac{1}{2}), \tag{11.16}$$

provided $w \in \Sigma_2(h)$, where

$$\Sigma_2(h) = \left\{ w = |w|e^{i\phi} : (1+h)\sigma(\phi) \leq |w| \leq 1, \ |\phi| \leq \pi/2\lambda \right\} . \tag{11.17}$$

Using (11.16) in (11.9) we find

$$(11.18)\qquad |U_m(w)| \le \exp\left(-\frac{m\lambda h^2}{4}\right) \qquad (m>m_o(h),\ 0<h\le\tfrac{1}{2},\ w\in\textstyle\sum_2(h)) .$$

Concerning the circular portion of the Szegö curve, let $h(0 < h < 1/2)$ be given and assume that

$$(11.19)\qquad \frac{\pi}{2\lambda} + h \le \arg w \le 2\pi - \left(\frac{\pi}{2\lambda} + h\right) .$$

Then from (4.6) and the asymptotic relations (6.2) and (6.3) (with $z = Rw$) we have that

$$(11.20)\qquad |U_m(w)| = \frac{|f(Rw)|}{a_m R^m |w|^m} = \frac{1+\eta}{R|w|^{m+1}\Gamma(1-\frac{1}{\lambda})\,a_m R^m} ,$$

for all $|Rw| \ge 2$, where

$$(11.21)\qquad \eta = \eta(w) = \frac{\omega K}{R|w|} \qquad (|\omega| \le 1) .$$

Taking logarithms in (11.20) and using (6.10) and (11.7), we obtain

$$(11.22)\qquad \log|U_m(w)| = \log(1+\eta) + \left\{\frac{1}{2} - \frac{1}{\lambda}\right\}\log\left(\frac{m}{\lambda}\right) + \frac{1}{2}\log 2\pi$$
$$- (m+1)\log|w| - \log\Gamma(1-\tfrac{1}{\lambda}) - \frac{m}{\lambda} - \frac{1}{2} + \frac{K\omega}{m}$$
$$(m > m_o ,\ |\omega| \le 1) .$$

Straight-forward upper estimates of the right-hand side of (11.22) gives, for $m > m_o(h)$,

$$(11.23)\qquad |U_m(w)| \le (1 + h)^{-m/2} \qquad (\lambda \ge 2,\ \ w\in\textstyle\sum_3(h)) ,$$

where

$$(11.24)\qquad \textstyle\sum_3(h) = \left\{w=|w|e^{i\phi} : (1+h)e^{-1/\lambda}\le|w|\le 1,\ \frac{\pi}{2\lambda}+h\le\phi\le 2\pi-\left(\frac{\pi}{2\lambda}+h\right)\right\} ,$$

and

$$(11.25)\qquad |U_m(w)| \le m^{(\lambda-2)/4\lambda} \qquad (1 < \lambda < 2,\ w \in \Sigma_4(h))\ ,$$

where

$$(11.26)\qquad \Sigma_4(h) = \left\{w=|w|e^{i\phi}:\ e^{-1/\lambda} \le |w| \le 1,\ \frac{\pi}{2\lambda} + h \le \phi \le 2\pi - \left(\frac{\pi}{2\lambda} + h\right)\right\}.$$

Similarly, if the positive constant $\tilde{B}_o = \tilde{B}_o(h,\lambda) \ge 2$ is chosen so that $|Rw| \ge \tilde{B}_o$ implies that, in (11.21), $|\eta| \le 1/2$, then equation (11.22) leads to the following lower estimates for $m > m_o(h)$:

$$(11.27)\qquad |U_m(w)| \ge (1 + h)^m \qquad (1 < \lambda < 2,\ w \in \Sigma_5(m,h))\ ,$$

where

$$(11.28)\qquad \Sigma_5(m,h) = \left\{w=|w|e^{i\phi}:\ \frac{\tilde{B}_o}{R_m} \le |w| \le (1-h)e^{-1/\lambda},\ \frac{\pi}{2\lambda} + h \le \phi \le 2\pi - \left(\frac{\pi}{2\lambda} + h\right)\right\};$$

$$(11.29)\qquad |U_m(w)| \ge m^{(\lambda-2)/4\lambda} \qquad (\lambda > 2\ ,\ w \in \Sigma_6(m,h))\ ,$$

where

$$(11.30)\qquad \Sigma_6(m,h) = \left\{w=|w|e^{i\phi}:\ \frac{\tilde{B}_o}{R_m} \le |w| \le e^{-1/\lambda},\ \frac{\pi}{2\lambda} + h \le \phi \le 2\pi - \left(\frac{\pi}{2\lambda} + h\right)\right\}.$$

To treat the omitted case $\lambda = 2$, we choose the constant $\tilde{B}_o$ so that $|Rw| \ge \tilde{B}_o$ implies

$$|\eta| \le 1 - 2^{-1/7}\ .$$

Then, for $\tilde{B}_o/R \le |w| \le e^{-1/2}$, a more careful analysis of (11.22) gives

$$\log|U_m(w)| \ge -\frac{1}{7}\log 2 + \frac{1}{2}\log 2\pi + \frac{m+1}{2} - \log \Gamma\left(\frac{1}{2}\right)$$
$$-\frac{m}{2} - \frac{1}{2} - \frac{K}{m} = \frac{5}{14}\log 2 - \frac{K}{m}\ .$$

Thus, we have

$$(11.31)\qquad |U_m(w)| \ge 2^{1/3} \qquad (\lambda = 2\ ,\quad w \in \Sigma_7(m,h)),$$

where

$$(11.32)\quad \Sigma_7(m,h) = \left\{w=|w|e^{i\phi}:\ \frac{\tilde{B}_o}{R_m}\le|w|\le e^{-1/2}\ ,\ \frac{\pi}{4}+h\le\phi\le 2\pi-\left(\frac{\pi}{4}+h\right)\right\}\ .$$

In order to pass from the estimates for $U_m(w)$ to those for $Q_m(w)$ we use (4.13), (7.11) and Lemma 9.2.

<u>Lemma 11.1</u>. <u>Let</u> $\xi = \sigma(\phi)e^{i\phi}$ <u>be any point of the normalized Szegö curve and let</u> $h\,(0 < h < 1/2)$ <u>be given. Then on the sets</u> $\Sigma_1, \Sigma_2, \ldots, \Sigma_7$ <u>defined in</u> (11.11), (11.17), (11.24), (11.26), (11.28), (11.30) <u>and</u> (11.32) <u>there holds</u>:

$$(11.33)\quad |Q_m(w)| \ge \frac{(2-\sqrt{3})}{2\sqrt{6}}(\pi m\lambda)^{1/2} > 0 \qquad (w\in\Sigma_1(m),\ m > m_o)\ ,$$

$$(11.34)\quad |Q_m(w)| \ge \frac{e^{-1/\lambda}}{3} \qquad (w\in\Sigma_2(h),\ m > m_o(h)),$$

$$(11.35)\quad |Q_m(w)| \ge \frac{e^{-1/\lambda}}{3} \qquad (\lambda \ge 2,\ w\in\Sigma_3(h),\ m > m_o(h)),$$

$$(11.36)\quad |Q_m(w)| \ge \frac{e^{-1/\lambda}}{3} \qquad (1 < \lambda < 2,\ w\in\Sigma_4(h),\ m > m_o(h))\ ,$$

$$(11.37)\quad |Q_m(w)| \ge \frac{1}{2}(1+h)^m \qquad (1 < \lambda < 2,\ w\in\Sigma_5(m,h),\ m > m_o(h))\ ,$$

$$(11.38)\quad |Q_m(w)| \ge \frac{1}{2}\, m^{(\lambda-2)/4\lambda} \qquad (\lambda > 2,\ w\in\Sigma_6(m,h),\ m > m_o(h))\ ,$$

$$(11.39)\quad |Q_m(w)| \ge 2^{1/3}-1 \qquad (\lambda = 2,\ w\in\Sigma_7(m,h),\ m > m_o(h))\ .$$

<u>Proof</u>: To prove (11.33) we note that (4.13) implies

$$(11.40)\qquad |Q_m(w)| \ge |U_m(w)| - |G_m(w)|\ .$$

In view of (4.8) and (4.10)

$$|G_m(w)| \le G_m(1) \qquad (|w| \le 1)\ ;$$

hence, for $w \in \Sigma_1(m)$,

$$(11.41) \qquad |Q_m(w)| \geq |U_m(w)| - G_m(1) .$$

Now, from (7.7) (with $\alpha = 3$), (7.11), and (8.17), it follows that

$$(11.42) \qquad G_m(1) = \left(\frac{\pi m\lambda}{2}\right)^{1/2} + K\omega , \quad (m > m_o,\ |\omega| \leq 1) ,$$

and so from (11.10) and (11.41) we find for $w \in \Sigma_1(m)$

$$(11.43) \quad |Q_m(w)| \geq \left(\frac{2\pi m\lambda}{3}\right)^{1/2} - \left(\frac{\pi m\lambda}{2}\right)^{1/2} - K\omega \geq \frac{(2-\sqrt{3})(\pi m\lambda)^{1/2}}{2\sqrt{6}} , \ (m>m_o) .$$

For $w \in \Sigma_5$, Σ_6 or Σ_7, the convergence relation (4.12) implies that the functions $G_m(w)$ are uniformly bounded on these sets. Hence (11.37) and (11.38) are immediate consequences of (11.40), (11.27), and (11.29). To establish (11.39), we note that for $|w| \leq e^{-1/2}$ and $\pi/4 \leq \arg w \leq 7\pi/4$, there holds

$$(11.44) \quad \left|\frac{w}{1-w}\right| \leq \frac{e^{-1/2}}{\{1-2|w|\cos\phi+|w|^2\}^{1/2}} \leq \frac{e^{-1/2}}{\{1-\sqrt{2}e^{-1/2}+e^{-1}\}^{1/2}} < 0.8493 < 1 .$$

Thus, as $G_m(w) \to w/(1-w)$ uniformly for $|w| \leq e^{-1/2}$, the estimate (11.39) follows from (11.31), (11.40) and (11.44).

Finally, to prove (11.34), (11.35) and (11.36) we write

$$(11.45) \qquad |Q_m(w)| \geq |G_m(w)| - |U_m(w)|$$

instead of (11.40). The distance δ between the point 1 and any of the sets Σ_2, Σ_3, Σ_4 is clearly positive and only depends on the positive parameters h and λ; its value is immaterial. Hence, by Lemma 9.2,

$$(11.46) \quad |G_m(w)| > 2e^{-1/\lambda}/5 , \ (w \in \{\Sigma_2 \cup \Sigma_3 \cup \Sigma_4\},\ m > m_o(h)) .$$

It is now obvious that (11.34), (11.35) and (11.36) follow from (11.46), (11.18), (11.23) and (11.25).

12. Proof of Theorem 5. Assertion I of the theorem is an immediate consequence of the Eneström-Kakeya theorem which, in view of (6.12), may be applied to the second form of $Q_m(w)$ in (4.7).

Returning to the variable

$$z = R_m w ,$$

we may rewrite (4.7) as

$$s_m(z) = a_m z^m Q_m(z/R_m) ;$$

hence assertions II and III (ii) follow from Lemma 11.1. With regard to B(h) we may take it to be any positive quantity satisfying (2.43) and

$$B(h) > \max\{B_o, \tilde{B}_o\}.$$

Assertion III (i) of Theorem 5 follows from Hurwitz' theorem and the uniform convergence $s_m(z) \to E_{1/\lambda}(z)$ $(|z| \leq B)$.

13. Proof of Theorem 1. By (8.28) of Lemma 8.1,

$$(13.1) \quad \left(\frac{2}{\pi\lambda m}\right)^{1/2} Q_m\left(\frac{1}{1-\left(\frac{2}{\lambda m}\right)^{1/2}\zeta}\right) \to e^{\zeta^2}\left(1-\frac{2}{\pi^{1/2}}\int_0^{\zeta} e^{-t^2}\,dt\right)$$

$$= e^{\zeta^2}\operatorname{erfc}(\zeta) \qquad (m \to +\infty)$$

uniformly on every compact subset of the ζ-plane and, in particular, in the disk

$$(13.2) \qquad |\zeta| \leq t \qquad (t > 0) .$$

Define $\zeta_{1,m}$ by the relation

$$(13.3) \qquad \frac{1}{1-\left(\frac{2}{\lambda m}\right)^{1/2}\zeta_{1,m}} = 1+\left(\frac{2}{\lambda m}\right)^{1/2}\zeta$$

so that

$$(13.4) \qquad |\zeta-\zeta_{1,m}| \le \left(\frac{2}{\lambda m}\right)^{1/2} t(t+1) \qquad (m > m_o) .$$

Since the convergence in (13.1) is also uniform on the disk

$$(13.5) \qquad |\zeta| \le t+1 ,$$

we have, by (13.1), (13.3), (13.4) and (13.5)

$$(13.6) \qquad \left(\frac{2}{\pi\lambda m}\right)^{1/2} \left\{ Q_m\left(\frac{1}{1-\left(\frac{2}{\lambda m}\right)^{1/2}\zeta}\right) - Q_m\left(1+\left(\frac{2}{\lambda m}\right)^{1/2}\zeta\right)\right\} =$$

$$= \exp(\zeta^2)\ \mathrm{erfc}(\zeta) - \exp(\zeta_{1,m}^2)\ \mathrm{erfc}(\zeta_{1,m}) + \eta(\zeta, \zeta_{1,m}) ,$$

where $\eta(\zeta, \zeta_{1,m}) \to 0$ $(m \to +\infty)$, uniformly for all ζ restricted by (13.2).

From (13.4) we deduce that, in (13.6),

$$|\exp(\zeta^2)\ \mathrm{erfc}(\zeta) - \exp(\zeta_{1,m}^2)\ \mathrm{erfc}(\zeta_{1,m})| \le$$

$$|\zeta-\zeta_{1,m}| \max_{|\zeta|\le t+1} \left|\frac{d}{d\zeta}\left(\exp(\zeta^2)\ \mathrm{erfc}(\zeta)\right)\right| \le K m^{-\frac{1}{2}} .$$

Hence (13.6) and (13.1) imply

$$(13.7) \qquad \left(\frac{2}{\pi\lambda m}\right)^{1/2} Q_m\left(1+\left(\frac{2}{\lambda m}\right)^{1/2}\zeta\right) \to \exp(\zeta^2)\ \mathrm{erfc}(\zeta) ,$$

uniformly on every compact set of the ζ-plane.

By (6.10) and (6.2)

$$(13.8) \qquad a_m R^m \sim f(R)(2\pi\lambda m)^{-\frac{1}{2}} \qquad (m \to \infty,\ R = R_m) .$$

Using (4.7) to return to the sections $s_m(z)$, we see that (13.7) and (13.8) yield, as $m \to \infty$,

$$(13.9) \qquad \left(1+\left(\frac{2}{\lambda m}\right)^{1/2}\zeta\right)^{-m}\{f(R_m)\}^{-1} s_m\left(R_m\left(1+\left(\frac{2}{\lambda m}\right)^{1/2}\zeta\right)\right) \to \frac{1}{2}\exp(\zeta^2)\ \mathrm{erfc}(\zeta) ,$$

uniformly on every compact set of the ζ-plane.

This limiting relation coincides with (2.13) which is thus proved.

Let $\zeta_j (j = 1,2,3,\ldots,\nu)$ be all the zeros (necessarily simple) of $\mathrm{erfc}(\zeta)$, which lie in the disk (13.2). (We have assumed that there are no zeros on the circumference $|\zeta| = t$.)

By Hurwitz' theorem, if $m > m_o$, the zeros of the left-hand side of (13.9), and therefore also the zeros of

$$(13.10) \qquad s_m\left(R_m\left(1+\left(\frac{2}{\lambda m}\right)^{1/2}\zeta\right)\right) ,$$

are all simple and "close" to the quantities ζ_j . Hurwitz' theorem also asserts that the functions in (13.10) have no other zeros in the disk $|\zeta| \leq t$. It is now obvious that the preceding argument leads at once to the precise formulation of our Theorem 1.

14. Proof of Theorem 2. Consider the point $\xi = \xi(\phi)$ $(0 < \phi \leq \frac{\pi}{2\lambda})$ of the normalized Szegö curve. Its definition, contained in (10.2) and (10.3) implies

$$(14.1) \qquad 0 = X(|\xi|) = |\xi|^{\lambda}\cos(\phi\lambda) - 1 - \lambda \log |\xi| ,$$

and hence

$$\xi^{\lambda} - 1 - \lambda \log \xi = i(|\xi|^{\lambda}\sin(\lambda\phi) - \lambda\phi) = i\tau.$$

With τ thus defined we introduce a real sequence $(\tau_m)_m$ by the conditions

$$(14.2) \qquad -\pi < \tau_m \leq \pi , \qquad \frac{\tau}{\lambda} m \equiv \tau_m \qquad (\mathrm{mod}\ 2\pi) .$$

Relevant information regarding our problem will now be derived from the study of

$$Q_m(\xi u)$$

where u is a complex variable which, for reasons that will become obvious, we rewrite in the form

$$(14.3) \qquad u = 1 + \frac{\log m}{2(1-\xi^{\lambda})m} - \frac{\zeta - i\tau_m}{(1-\xi^{\lambda})m} \qquad (\zeta = \rho e^{i\theta}).$$

Our proof is obtained by studying the sequence of polynomials $\{T_m(\zeta)\}_m$ where

$$(14.4) \qquad T_m(\zeta) = Q_m\left(\xi\left(1 + \frac{\log m}{2(1-\xi^{\lambda})m} - \frac{\zeta - i\tau_m}{(1-\xi^{\lambda})m}\right)\right) \qquad (m = 1,2,3,\ldots).$$

Under the additional restrictions

$$(14.5) \qquad 0 \leq \rho \leq B, \qquad m > m_o(\phi,B),$$

we find, in view of (14.2),

(14.6) $$\frac{1}{2}e^{-1/\lambda} < |\xi u| < 2, \qquad |\arg(\xi u)| < \frac{\pi}{\lambda}\left(\frac{1}{2}+\frac{1}{5}\right).$$

This enables us to use the asymptotic representation (6.2), which now yields

(14.7) $$f(R\xi u) = \lambda\ \exp(R^\lambda\xi^\lambda u^\lambda) - \frac{1}{c_1 R\xi u} + \frac{\omega K}{R^2} \qquad \left(c_1 = \Gamma\left(1-\frac{1}{\lambda}\right)\right).$$

In order to evaluate $U_m(R\xi u)$ (U_m is defined in (4.6)), we remark that (6.10) implies

(14.8) $$\log\left(a_m R^m \xi^m u^m\right) = \frac{m}{\lambda} - \frac{1}{2}\log\left(\frac{m}{\lambda}\right) + \frac{1}{2} - \frac{1}{2}\log 2\pi + \frac{\omega K}{m} + m\log|\xi|$$

$$+ mi\phi + m\log u\ ,$$

where, by (14.3) and (14.5)

(14.9) $$m\log u = \frac{\log m}{2(1-\xi^\lambda)} - \frac{\zeta - i\tau_m}{1-\xi^\lambda} + \frac{\omega K(\log\ m)^2}{m}.$$

Hence

(14.10) $$a_m R^m \xi^m u^m = \left(\frac{e\lambda}{2\pi}\right)^{1/2} \exp\left\{\left(\frac{1}{\lambda} + \log|\xi|\right)m + \frac{\xi^\lambda}{2(1-\xi^\lambda)}\log m - \frac{\zeta - i\tau_m}{1-\xi^\lambda}\right.$$

$$\left. + \ im\phi + \frac{\omega K(\log\ m)^2}{m}\right\}.$$

If

(14.11) $$0 < \phi < \frac{\pi}{2\lambda},$$

then (14.1) yields

(14.12) $$\frac{1}{\lambda} + \log|\xi| = \frac{|\xi|^\lambda\cos(\phi\lambda)}{\lambda} > 0\ ,$$

and hence (14.10) implies (uniformly)

$$(14.13)\qquad |a_m R^m \xi^m u^m| > \exp(c_2 m) \qquad (c_2 > 0) ,$$

where we may choose $c_2 = |\xi|^\lambda \cos(\phi\lambda)/2\lambda$.

If

$$(14.14)\qquad \phi = \frac{\pi}{2\lambda} ,$$

then, the left-hand side of (14.12) vanishes, and (14.10) only yields

$$(14.15)\qquad |a_m R^m \xi^m u^m| > K \exp\left(-\frac{1}{2(1+e^2)} \log m\right) \qquad (m > m_o) .$$

Write

$$(14.16)\quad \Omega = R^\lambda \xi^\lambda u^\lambda - \left(\frac{1}{\lambda} + \log |\xi|\right)m - im\phi - \frac{\xi^\lambda}{2(1-\xi^\lambda)} \log m + \frac{\zeta - i\tau_m}{1-\xi^\lambda} .$$

Then, by (14.7), (14.13) and (4.6), we find

$$(14.17)\qquad \frac{f(R\xi u)}{a_m R^m \xi^m u^m} = U_m(\xi u) = \left(\frac{2\pi\lambda}{e}\right)^{1/2} \exp\left(\Omega + \frac{\omega K(\log m)^2}{m}\right) + \omega K \exp(-c_2 m) ,$$

provided (14.11) holds.

If we assume (14.14) instead of (14.11), the form of (14.17) is modified; taking (14.15) and (4.5) into account, and restricting λ by the condition (2.20) we obtain

$$(14.18)\quad U_m(\xi u) = \left(\frac{2\pi\lambda}{e}\right)^{1/2} \exp\left(\Omega + \frac{\omega K(\log m)^2}{m}\right) + \omega K \exp\left\{\left(\frac{1}{2(1+e^2)} - \frac{1}{\lambda}\right)\log m\right\}$$

If

$$2(1+e^2) \leq \lambda ,$$

the error term in (14.18) is not negligible. In this case, the behavior of $U_m(\xi u)$ requires a closer study, which will be found in §16.

To simplify Ω we use (11.7), (14.3) and (14.2). After some straightforward reductions we find

$$\Omega = \zeta + \frac{\xi^\lambda}{2} + i2\pi q + \frac{\omega K(\log m)^2}{m} \qquad (q = \text{integer}),$$

and hence, (14.17) yields

$$(14.19) \qquad U_m(\xi u) \to (2\pi\lambda)^{1/2}\exp\left(\frac{1}{2}(\xi^\lambda - 1)\right)e^\zeta, \quad (m \to +\infty),$$

uniformly on every compact subset of the ζ-plane. If $\phi = \pi/2\lambda$, then (14.18) and the additional assumption of (2.20) lead, once more, to (14.19).

Now by (4.13) and (14.4),

$$(14.20) \qquad T_m(\zeta) = U_m(\xi u) - G_m(\xi u),$$

and since $|\xi| < 1$, we have by (4.12),

$$(14.21) \qquad \lim_{m \to \infty} G_m(\xi u) = \frac{\xi}{1-\xi},$$

where the limit holds uniformly for $|\zeta| = \rho \leq B$.

From (14.19), (14.20) and (14.21) we conclude that

$$(14.22) \quad T_m(\zeta) \to (2\pi\lambda)^{1/2}\exp\left(\frac{1}{2}(\xi^\lambda - 1)\right)e^\zeta - \frac{\xi}{1-\xi} \qquad (m \to +\infty),$$

uniformly on every compact subset of the ζ-plane.

The limit function in (14.22) has simple zeros at the points

$$(14.23) \quad \zeta_k = -\frac{1}{2}\log(2\pi\lambda) - \frac{1}{2}(\xi^\lambda - 1) + \log\left(\frac{\xi}{1-\xi}\right) + 2k\pi i \qquad (k=0,\pm1,\pm2,\ldots).$$

To determine ζ_o without ambiguity we use (14.23) with $k = 0$ and select the determination of the logarithms such that

$$-\pi < \mathrm{Im}\left\{\log\left(\frac{\xi}{1-\xi}\right) - \frac{1}{2}\log(2\pi\lambda)\right\} \leq \pi .$$

By Hurwitz' theorem, $T_m(\zeta)$ has simple zeros $\zeta_o + \zeta_{m,k}$ $(k=0,\pm1,\pm2,\ldots)$, which in view of our notational convention regarding η_m, are located at the points

$$\zeta_o + \zeta_{m,k} = \zeta_o + 2k\pi i + \eta_m(k) \quad (k=0,\pm1,\pm2,\ldots) . \tag{14.24}$$

We now return to the z-plane and state our results so as to exhibit the zeros of the sections $s_m(z)$ of $E_{1/\lambda}(z)$. By (4.7), (14.4) and (14.24) we find

$$0 = T_m(\zeta_o+\zeta_{m,k}) = s_m\left(R\xi\left(1 + \frac{\log m+2i\tau_m}{2(1-\xi^\lambda)m} - \frac{\zeta_o+2k\pi i+\eta_m}{(1-\xi^\lambda)m}\right)\right) \tag{14.25}$$

$$(k=0,\pm1,\pm2,\ldots) ,$$

where $\eta_m = \eta_m(k) \to 0$ as $m \to +\infty$ and τ_m is given by (14.2). It is now obvious that Theorem 2 follows at once from (14.25).

15. The circular portion of the Szegö curve (Proof of Theorem 3).

In the last section we studied the zeros of the normalized sections $s_m(R_m w)$ for w in a neighborhood of a point $\xi(\neq 1)$ of the Szegö curve in the sector $|\phi| \leq \pi/2\lambda$. The remainder of the Szegö curve is the circular arc

$$\xi = \exp(-\tfrac{1}{\lambda} + i\phi) \quad (\tfrac{\pi}{2\lambda} < \phi < 2\pi - \tfrac{\pi}{2\lambda}) . \tag{15.1}$$

In this section we characterize the zeros of $s_m(R_m w)$ for w near a fixed point ξ satisfying (15.1).

With the assumption of (15.1), introduce

$$\tilde{\xi} = \xi\exp\left\{\left(\frac{1}{2} - \frac{1}{\lambda}\right)\frac{\log m}{m+1}\right\} , \tag{15.2}$$

and, by analogy with (14.2), consider $\tilde{\tau}_m$ characterized by

$$(15.3)\qquad \tilde{\tau}_m \equiv (m+1)\phi \quad (\mathrm{mod}\ 2\pi), \quad -\pi < \tilde{\tau}_m \leq \pi .$$

Keeping m and ϕ fixed, we perform the change of variable

$$(15.4)\qquad w = \tilde{\xi}\exp\left(\frac{\tilde{\zeta}_o - i\tilde{\tau}_m}{m+1}\right)\left(1 + \frac{\zeta}{m+1}\right)$$

$$= \exp\left\{-\frac{1}{\lambda} + i\phi + \left(\frac{1}{2} - \frac{1}{\lambda}\right)\frac{\log m}{m+1} + \frac{\tilde{\zeta}_o - i\tilde{\tau}_m}{m+1}\right\}\left(1 + \frac{\zeta}{m+1}\right) \quad (\zeta = \rho e^{i\theta}),$$

where ζ is the new variable and $\tilde{\zeta}_o$ is defined by (2.31). From (4.13) we have

$$(15.5)\qquad Q_m\left(\tilde{\xi}\exp\left(\frac{\tilde{\zeta}_o - i\tilde{\tau}_m}{m+1}\right)\left(1 + \frac{\zeta}{m+1}\right)\right)$$

$$= U_m\left(\tilde{\xi}\exp\left(\frac{\tilde{\zeta}_o - i\tilde{\tau}_m}{m+1}\right)\left(1 + \frac{\zeta}{m+1}\right)\right) - G_m\left(\tilde{\xi}\exp\left(\frac{\tilde{\zeta}_o - i\tilde{\tau}_m}{m+1}\right)\left(1 + \frac{\zeta}{m+1}\right)\right) .$$

In order to simplify the notation, write U_m and G_m for the two expressions in the right-hand side of (15.5).

We now assume

$$(15.6)\qquad |\zeta| \leq B , \quad m > m_o(B,\phi) .$$

Then, from (4.12) and (15.4) we conclude that, as $m \to +\infty$,

$$(15.7)\qquad G_m = \frac{\xi}{1-\xi}(1 + \eta_m) \quad (\eta_m \to 0) ,$$

where the relation $\eta_m \to 0$ holds uniformly for all $|\zeta| \leq B$.

To evaluate U_m we note that, under the restriction (15.6), one of the two relations (6.2) or (6.3) implies

$$(15.8)\qquad f\left(R\tilde{\xi}\exp\left(\frac{\tilde{\zeta}_o - i\tilde{\tau}_m}{m+1}\right)\left(1 + \frac{\zeta}{m+1}\right)\right)$$

$$= -\left(1 + \frac{\omega K}{R}\right)\frac{1}{\Gamma\left(1 - \frac{1}{\lambda}\right)}\left\{R\tilde{\xi}\exp\left(\frac{\tilde{\zeta}_o - i\tilde{\tau}_m}{m+1}\right)\left(1 + \frac{\zeta}{m+1}\right)\right\}^{-1} ,$$

and hence, by (4.6),

$$(15.9)\qquad U_m = -\frac{\left(1+\frac{\omega K}{R}\right)\exp\left(-\tilde{\zeta}_o + i\tilde{\tau}_m\right)\left(1+\frac{\zeta}{m+1}\right)^{-m-1}}{R\tilde{\xi}^{m+1}\Gamma\left(1-\frac{1}{\lambda}\right)a_m R^m}$$

$$= -\frac{\left(1+\frac{\omega K}{R}\right)}{\Gamma\left(1-\frac{1}{\lambda}\right)}\left(1+\frac{\zeta}{m+1}\right)^{-m-1} e^{-\Lambda} \quad ,$$

where

$$\Lambda = (m+1)\log\tilde{\xi} + \log R + \log(a_m R^m) + \tilde{\zeta}_o - i\tilde{\tau}_m \ .$$

To simplify the form of Λ we use (15.2), (4.5), (2.30), (2.31) and (6.10); after some straightforward reductions we find

$$(15.10)\qquad \Lambda = \tilde{\zeta}_o + \left(\frac{1}{2} - \frac{1}{\lambda}\right)\log(e\lambda) - \frac{1}{2}\log(2\pi) + i2q\pi + \frac{\omega K}{m}$$

$$= -\log\Gamma\left(1-\frac{1}{\lambda}\right) + \log(e^{1/\lambda}e^{-i\phi}{}_{-1}) + i2q\pi + i\pi + \frac{\omega K}{m} \quad (q=\text{integer}).$$

Combining (15.9) and (15.10) we obtain

$$(15.11)\qquad U_m \to \frac{e^{-\zeta}}{e^{1/\lambda}e^{-i\phi}{}_{-1}} \qquad (m \to +\infty) \ ,$$

uniformly in any compact set of the ζ-plane. From (15.5), (15.7) and (15.11), we conclude that as $m \to +\infty$, we have

$$(15.12)\qquad Q_m\left(\tilde{\xi}\exp\left(\frac{\tilde{\zeta}_o - i\tilde{\tau}_m}{m+1}\right)\left(1+\frac{\zeta}{m+1}\right)\right) \to \frac{e^{-\zeta}-1}{e^{1/\lambda}e^{-i\phi}{}_{-1}} \ ,$$

uniformly in $|\zeta| \leq t$ $(t > 0)$.

Selecting $\tilde{p}_m$ as indicated in (2.32), we complete the proof of Theorem 3 exactly as the analogous proof of Theorem 2.

16. Proof of Theorem 4. It suffices to follow, step by step, the analogous proof of Theorem 3.

We first introduce

$$(16.1) \qquad \xi' = \exp\left(-\frac{1}{\lambda} + i\frac{\pi}{2\lambda} + \left\{\left(\frac{1}{2} - \frac{1}{\lambda}\right) + i\alpha\right\} \frac{\log m}{m+1}\right),$$

and, by analogy with (15.3) we also consider

$$(16.2) \qquad \tau'_m \equiv (m+1)\frac{\pi}{2\lambda} + \alpha \log m \quad (\mathrm{mod}\ 2\pi), \qquad -\pi < \tau'_m \leq \pi .$$

Keeping m fixed, we perform the change of variable

$$(16.3) \qquad w = \xi' \exp\left(\frac{\zeta'_o - i\tau'_m}{m+1}\right)\left(1 + \frac{\zeta}{m+1}\right) \qquad (\zeta = \rho e^{i\theta}),$$

where ζ is the new variable and ζ'_o is defined by (2.37).

Once more we restrict ζ and m by conditions such as

$$(16.4) \qquad |\zeta| \leq B, \qquad m > m_o = m_o(B) ,$$

and, by (16.3) and (4.13), we have

$$(16.5) \qquad Q_m\left(\xi' \exp\left(\frac{\zeta'_o - i\tau'_m}{m+1}\right)\left(1 + \frac{\zeta}{m+1}\right)\right)$$

$$= U_m\left(\xi' \exp\left(\frac{\zeta'_o - i\tau'_m}{m+1}\right)\left(1 + \frac{\zeta}{m+1}\right)\right) - G_m\left(\xi' \exp\left(\frac{\zeta'_o - i\tau'_m}{m+1}\right)\left(1 + \frac{\zeta}{m+1}\right)\right).$$

In order to simplify the notation, write U_m and G_m for the two expressions in the right-hand side of (16.5).

From (4.12) and (16.3) we conclude that, as $m \to +\infty$,

$$(16.6) \qquad G_m = \frac{\xi}{1-\xi}(1 + \eta_m) \quad \left(\xi = \exp\left(-\frac{1}{\lambda} + \frac{i\pi}{2\lambda}\right),\ \eta_m \to 0\right),$$

where the relation $\eta_m \to 0$ holds uniformly for all $|\zeta| \leq B$.

To evaluate U_m we note that, under the restriction (16.4), the relation (6.2) implies

$$(16.7)\qquad f\left(R\xi'\exp\left(\frac{\zeta_o' - i\tau_m'}{m+1}\right)\left(1 + \frac{\zeta}{m+1}\right)\right) = \lambda\exp\left((R\xi')^\lambda\left(1 + \frac{\omega K}{m+1}\right)\right)$$

$$- \frac{1}{\Gamma\left(1 - \frac{1}{\lambda}\right)}\left\{R\xi'\exp\left(\frac{\zeta_o' - i\tau_m'}{m+1}\right)\left(1 + \frac{\zeta}{m+1}\right)\right\}^{-1} + \frac{\omega K}{R^2} .$$

To estimate the first term in the right-hand side of (16.7) we note that (11.7) and (16.1) imply

$$0 < R^\lambda < \frac{2m}{\lambda}, \qquad |\xi'|^\lambda < 1 \qquad (m > m_o) ,$$

and hence, using again (11.7) and (16.1),

$$(R\xi')^\lambda\left(1 + \frac{\omega K}{m+1}\right) = (R\xi')^\lambda + \omega K = \frac{m}{\lambda}(\xi')^\lambda + \omega K$$

$$= \frac{m}{\lambda}e^{-1}i\ \exp\left\{\left(\frac{\lambda}{2} - 1\right)\frac{\log m}{m+1}\right\}\exp\left(i\alpha\lambda\frac{\log m}{m+1}\right) + \omega K ,$$

$$\operatorname{Re}\left\{(R\xi')^\lambda\left(1 + \frac{\omega K}{m+1}\right)\right\} \le -\alpha e^{-1}(1 + \eta_m)\log m + \omega K .$$

We have thus shown that

$$(16.8)\qquad \left|\lambda\exp\left((R\xi')^\lambda\left(1 + \frac{\omega K}{m+1}\right)\right)\right| < Km^{-\alpha e^{-1} + \eta_m} ,$$

uniformly for all ζ and m satisfying (16.4).

To complete our estimate of U_m, we use (6.10) and (16.3); this leads to

$$(16.9)\qquad \log\left(a_m R^m (\xi')^m \exp\left(\frac{m}{m+1}(\zeta_o' - i\tau_m')\right)\left(1 + \frac{\zeta}{m+1}\right)^m\right)$$

$$= -\frac{1}{2}\log m + \frac{1}{2}\log\left(\frac{e\lambda}{2\pi}\right) + \frac{\omega K}{m} + mi\frac{\pi}{2\lambda}$$

$$+ \frac{m \log m}{m+1}\left(\frac{1}{2} - \frac{1}{\lambda} + i\alpha\right) + \frac{m}{m+1}(\zeta_o' - i\tau_m') + m\log\left(1 + \frac{\zeta}{m+1}\right) .$$

In view of (2.35), (2.36), (4.6), (16.7), (16.8) and (16.9), we obtain

$$U_m = -\frac{1}{\Gamma\left(1 - \frac{1}{\lambda}\right)}\left(1 + \frac{\zeta}{m+1}\right)^{-(m+1)} \exp(-\Lambda') + \eta_m ,$$

where

$$\Lambda' = \zeta_o' + \left(\frac{1}{2} - \frac{1}{\lambda}\right)\log(e\lambda) - \frac{1}{2}\log(2\pi) + i2q\pi + \frac{\omega K}{m} \qquad (q = \text{integer}) .$$

This formula is entirely analogous to (15.10). Hence, by the arguments following (15.10) in §15, we find

$$U_m \to \frac{e^{-\zeta}}{e^{1/\lambda}e^{-i\pi/2\lambda} - 1} \qquad (m \to +\infty) ,$$

uniformly on any compact set of the ζ-plane.

It is now clear that the proof of Theorem 4 may be completed exactly as the analogous proof of Theorem 3.

17. <u>Proof of Theorem 6</u>. For each $t > 0$, and $m > m_0(t)$, it is easily seen that there is a unique point $\xi_{t,m}$ of the normalized Szegö curve S which lies in the upper half-plane and satisfies

(17.1) $$|\xi_{t,m} - 1| = t\left(\frac{2}{\lambda m}\right)^{1/2} .$$

Write

(17.2) $$\xi_{t,m} = 1 + \mu_{t,m} t\left(\frac{2}{\lambda m}\right)^{1/2}, \qquad |\mu_{t,m}| = 1.$$

The tangent line at w=1, to the upper portion of the symmetric curve S forms an angle $3\pi/4$ with the positive axis. Hence, for each fixed $t > 0$, we have

(17.3) $$\lim_{m \to \infty} \mu_{t,m} = \exp(i3\pi/4).$$

Consequently, there exists a $\hat{t} > 0$ such that

(17.4) $$\hat{t}\ \mathrm{Im}(\mu_{\hat{t},m}) > 2\ \mathrm{Im}(\zeta^*), \qquad (m > m_0(\hat{t})),$$

where ζ^* is the zero of $\mathrm{erfc}(\zeta)$ with smallest positive imaginary part. For later use we impose the additional restriction

(17.5) $$\mathrm{erfc}(\hat{t}e^{i\Theta}) \neq 0 \qquad (0 \leq \Theta < 2\pi);$$

this is clearly possible.

Next, we fix an angle ψ such that

(17.6) $$0 < \psi < \pi/2\lambda, \qquad \frac{3}{2}\left[\sigma(\psi)\cos\psi\right]^{\lambda/2} > 1,$$

where $\sigma(\psi)$ is the modulus of the point on S with argument ψ. Finally, select the positive constant B such that

(17.7) $$B^{\lambda/2}\tan\psi > \frac{\sqrt{2}}{\lambda}\ \mathrm{Im}(\zeta^*), \qquad B > \max(1, B(h)),$$

where $B(h)$ is the constant in Theorem 5 (III), with $h = \frac{1}{4}$.

Let

$$w = u + iv \qquad (u,\ v,\ \text{real}),$$

be the variable of the normalized plane.

We decompose the half-plane

(17.8) $$u \geq \frac{B}{R_m}$$

in three disjoint sets D_1, D_2, D_3 (these sets depend on m) which are defined as follows:

(17.9) $$D_1 = D_1(m) = \{w: |w-1| \leq \hat{t}\left(\frac{2}{\lambda m}\right)^{1/2}\},$$

(17.10) $$D_2 = D_2(m) = \{w: |w-1| > \hat{t}\left(\frac{2}{\lambda m}\right)^{1/2}, \ |\arg w| \leq \psi, \text{ and } u \geq \frac{B}{R_m}\},$$

(17.11) $$D_3 = D_3(m) = \{w = u + iv: u \geq B/R_m \text{ and } \psi < \arg w < 2\pi - \psi\}.$$

Now suppose that $\tilde{w}$ satisfies

(17.12) $$s_m(R_m \tilde{w}) = 0, \quad \tilde{w} = \tilde{u} + i\tilde{v} = |\tilde{w}|\exp(i\tilde{\phi}), \quad \tilde{u} \geq B/R_m.$$

Observe that

(17.13) $$\frac{|\tilde{v}|}{\tilde{u}^{1-(\lambda/2)}} = \tilde{u}^{\lambda/2} \ |\tan \tilde{\phi}|,$$

and that $\tilde{w}$ is a member of one (and only one) of the sets D_1, D_2, D_3.

If $\tilde{w} \in D_1$, then, trivially

(17.14) $$\tilde{u} \geq 1 - \hat{t}\left(\frac{2}{\lambda m}\right)^{1/2}.$$

Furthermore, by Theorem 1, we have for $m > m_0(\hat{t})$

(17.15) $$\tilde{w} = 1 + \left(\frac{2}{\lambda m}\right)^{1/2} (\zeta_j + \eta_{m,j}),$$

for some $j=1,2,\ldots,\nu$, where $\zeta_1,\ldots,\zeta_\nu$ $(\nu = \nu(\hat{t}))$ are the zeros of $\operatorname{erfc}(\zeta)$ in the disk $|\zeta| \leq \hat{t}$. By (17.5) we know that $\operatorname{erfc}(\zeta) \neq 0$ on the circle $|\zeta| = \hat{t}$. Hence

(17.16) $$|\tan \tilde{\phi}| = \frac{|\tilde{v}|}{\tilde{u}} = \frac{\left(\frac{2}{\lambda m}\right)^{1/2} |\operatorname{Im}(\zeta_j + \eta_{m,j})|}{1 + \left(\frac{2}{\lambda m}\right)^{1/2} \operatorname{Re}(\zeta_j + \eta_{m,j})},$$

and so for the constant K satisfying (2.47), it follows from (17.13), (17.14), (17.16) and the definition of ζ^* that

$$\frac{|\tilde{v}|}{\tilde{u}^{1-(\lambda/2)}} > \left(\frac{2}{\lambda m}\right)^{1/2} \frac{\lambda K}{\sqrt{2}} = \left(\frac{\lambda}{m}\right)^{1/2} K \qquad (m > m_0(K,\hat{t})). \tag{17.17}$$

If $\tilde{w} \in D_2$, then, in addition to the conditions

$$|\tilde{w}-1| > \hat{t}\left(\frac{2}{\lambda m}\right)^{1/2}, \quad |\arg \tilde{w}| \leq \psi, \quad \tilde{u} \geq B/R_m, \tag{17.18}$$

it follows from Theorem 5 (I and III(ii)) that for $m > m_0$

$$\sigma(\tilde{\phi}) < |\tilde{w}| < 1 \qquad (\tilde{\phi} = \arg \tilde{w}). \tag{17.19}$$

It is then geometrically evident from (17.3) that

$$|\tan \tilde{\phi}| \geq \tan(\arg \xi_{\hat{t},m}) = \frac{\left(\frac{2}{\lambda m}\right)^{1/2} \hat{t}\ \mathrm{Im}(\mu_{\hat{t},m})}{1+\hat{t}\left(\frac{2}{\lambda m}\right)^{1/2} \mathrm{Re}(\mu_{\hat{t},m})}. \tag{17.20}$$

Hence, from (17.4), we have

$$|\tan \tilde{\phi}| \geq \frac{3}{2}\left(\frac{2}{\lambda m}\right)^{1/2} \mathrm{Im}(\zeta^*) \qquad (m > m_0). \tag{17.21}$$

Also, from the definition of S, given in (2.6), it is easy to show that $\sigma(\phi)\cos\phi = \mathrm{Re}(\xi(\phi))$ is a decreasing function of ϕ for $0 \leq \phi \leq \pi/2\lambda$. Thus from (17.18) and (17.19) we have

$$\tilde{u} = |\tilde{w}|\cos\tilde{\phi} \geq \sigma(\tilde{\phi})\cos\tilde{\phi} \geq \sigma(\psi)\cos\psi. \tag{17.22}$$

The inequalities (2.47), (17.6), (17.21) and (17.22) together with equation (17.13) then give

$$\begin{aligned}\frac{|\tilde{v}|}{\tilde{u}^{1-(\lambda/2)}} &\geq [\sigma(\psi)\cos\psi]^{\lambda/2}\cdot\frac{3}{2}\left(\frac{2}{\lambda m}\right)^{1/2} \mathrm{Im}(\zeta^*) \\ &> \left(\frac{2}{\lambda m}\right)^{1/2} \mathrm{Im}(\zeta^*) > \left(\frac{\lambda}{m}\right)^{1/2} K \qquad (m > m_0).\end{aligned} \tag{17.23}$$

If $\tilde{w} \in D_3$, then, from (17.7), (17.11) and (2.47), we have

$$(17.24) \qquad \frac{|\tilde{v}|}{\tilde{u}^{1-(\lambda/2)}} = \tilde{u}^{\lambda/2}|\tan\tilde{\phi}| \geq \left(\frac{B}{R_m}\right)^{\lambda/2} \tan\psi$$

$$> R_m^{-\lambda/2}\frac{\sqrt{2}}{\lambda}\,\mathrm{Im}(\zeta^*) \geq R_m^{-\lambda/2} K \qquad (m > m_0).$$

Return to the variable $z = R_m w$ and assume that

$$(17.25) \qquad s_m(\tilde{z}) = 0, \qquad \tilde{z} = \tilde{x} + i\tilde{y}, \qquad \tilde{x} \geq B.$$

Then from one of the relations (17.17), (17.23), or (17.24) (with $\tilde{z} = R_m\tilde{w}$), and in view of

$$R_m^{\lambda} > \frac{m}{\lambda},$$

(a consequence of (2.2)), we obtain

$$(17.26) \qquad \frac{|\tilde{y}|}{\tilde{x}^{1-(\lambda/2)}} = R_m^{\lambda/2}\frac{|\tilde{v}|}{\tilde{u}^{1-(\lambda/2)}} > K \qquad (m > m_0).$$

This completes the proof of the first part of Theorem 6 since $x_0\,(>B)$ can be chosen greater than the real parts of the zeros of finitely many of the partial sums.

To verify the sharpness of the assertion regarding the parabolic region (2.48) one may reason as follows.

By Theorem 1 define a sequence $\{z_m\}_m$ satisfying the relations

$$0 = s_m(z_m) = s_m(R_m w_m), \qquad z_m = R_m w_m = x_m + iy_m,$$

with

$$w_m = 1 + \left(\frac{2}{\lambda m}\right)^{1/2}(\zeta^* + \eta_m) \qquad (\eta_m \to 0,\ m \to +\infty).$$

Hence, as $m \to +\infty$,

$$\frac{y_m}{x_m} = \left(\frac{2}{\lambda m}\right)^{1/2}(\mathrm{Im}(\zeta^*) + \eta_m) \qquad (\eta_m \to 0),$$

with

$$x_m \sim R_m \sim \left(\frac{m}{\lambda}\right)^{1/\lambda} \quad ;$$

(2.49) clearly follows.

If we now replace, in (2.48), K by any constant

$$K' > \frac{\sqrt{2}}{\lambda}\, \mathrm{Im}(\zeta^*) \quad ,$$

(2.49) shows that, for m large enough, the zeros z_m will fall in the "enlarged" parabolic region thus defined.

The sharp character of Theorem 6 is now obvious.

18. Properties of $\mathcal{L}$-functions; proof of assertion I of Theorem 7.

There is no novelty in the results of this section; in slightly different forms, they are scattered throughout the mathematical literature. For the convenience of the reader, we have summarized them here and stated them in terms of our definitions and notations.

For sake of completeness we have sketched a proof of the fact that our $\mathcal{L}$-functions of genus zero may be represented, in the complex plane, by the asymptotic relation stated below as (18.17).

Let F(z) be given by (2.53). Obviously

(18.1) $$a_j > 0 \qquad (j = 0,1,2,3,\ldots)$$

and hence

(18.2) $$\log F(r) = \log M(r) \qquad (r > 0,\ M(r) = \max_{|z|=r} |F(z)|).$$

Writing

(18.3) $$a(r) = r\,\frac{F'(r)}{F(r)} \,,$$

we have $a(r) > 0$ $(r > 0)$ and by Hadamard's convexity theorem (three circle theorem)

$$a'(r) \geq 0 \qquad (r \geq 0). \tag{18.4}$$

Introduce the counting function of the zeros of $F(z)$

$$n(t) = \sum_{x_k \leq t} 1 \qquad (t \geq 0) \tag{18.5}$$

and consider the well-known representation of Valiron (cf. [36, p. 271])

$$\log F(z) = z\int_0^{+\infty} \frac{n(t)}{t(t+z)}\, dt \qquad (|\arg z| < \pi). \tag{18.6}$$

We take for granted

<u>Valiron's tauberian theorem</u> [37, p. 237]. <u>The asymptotic relation (2.54) implies</u>

$$n(r) \sim \frac{B_1 \sin \pi\lambda}{\pi}\, r^{\lambda} \qquad (r \to +\infty). \tag{18.7}$$

Let ε $(0 < \varepsilon < \frac{1}{2})$ be given and let $r_0(\varepsilon) = r_0 > 1 + x_1$ be such that $r > r_0$ implies

$$n(r) = \frac{B_1 \sin \pi\lambda}{\pi}\, r^{\lambda}(1 + \eta(r)), \qquad |\eta(r)| \leq \varepsilon . \tag{18.8}$$

Using (18.8) in (18.6), noticing that $n(t) \equiv 0$ $(0 \leq t < x_1)$, and writing

$$\gamma = \frac{B_1 \sin \pi\lambda}{\pi},$$

we find

$$\log F(z) = z\int_{r_0}^{+\infty} \frac{\gamma t^{\lambda}}{t(t+z)}\, dt + \omega\gamma|z|\varepsilon\int_{r_0}^{+\infty} \frac{t^{\lambda}}{t|t+z|}\, dt + z\int_{x_1}^{r_0} \frac{n(t)}{t(t+z)}\, dt. \tag{18.9}$$

In the above relation, confine z to the sector

$$\Delta = \Delta(\varepsilon_1) = \{z=re^{i\theta}: |\theta| \le \pi-\varepsilon_1, r > 0\} \quad , \tag{18.10}$$

where $0 < \varepsilon_1 < \pi$ and ε_1 is otherwise arbitrary. Then

$$|t+z| = |te^{-i\frac{\theta}{2}} + re^{i\frac{\theta}{2}}| \ge (t+r)\cos\frac{\theta}{2} \ge (t+r)\sin\frac{\varepsilon_1}{2} = (t+r)\gamma_1. \tag{18.11}$$

Now

$$\int_0^\infty \frac{t^{\lambda-1}}{t+z}\,dt = \frac{\pi}{\sin\pi\lambda} z^{\lambda-1} \qquad (|\arg z| < \pi,\ 0 < \lambda < 1), \tag{18.12}$$

is a well-known consequence of the elements of contour integration. Hence for the second integral in the right-hand side of (18.9) we have from (18.11) and (18.12)

$$\gamma|z| \int_{r_0}^{+\infty} \frac{t^\lambda}{t|t+z|}\,dt \le \frac{\gamma}{\gamma_1} \frac{\pi}{\sin\pi\lambda} r^\lambda = \frac{B_1}{\gamma_1} r^\lambda. \tag{18.13}$$

Similarly, $z \in \Delta$ implies

$$\left| z\int_{x_1}^{r_0} \frac{n(t)}{t(t+z)}\,dt \right| \le \frac{r}{\gamma_1} \int_{x_1}^{r_0} \frac{n(r_0)}{t(t+r)}\,dt < \frac{n(r_0)}{\gamma_1} \log\left(\frac{r_0}{x_1}\right), \tag{18.14}$$

as well as

$$\left| z\int_0^{r_0} \frac{t^\lambda}{t(t+z)}\,dt \right| \le \frac{r}{\gamma_1} \int_0^{r_0} \frac{t^{\lambda-1}}{t+r}\,dt < \frac{r_0^{\lambda}}{\lambda\gamma_1}. \tag{18.15}$$

Combining (18.9) and (18.12) - (18.15) we obtain

$$\log F(z) = B_1 z^\lambda + E(z), \tag{18.16}$$

where

$$|E(z)| \le \frac{\varepsilon B_1}{\gamma_1} r^\lambda + \frac{n(r_0)}{\gamma_1} \log\left(\frac{r_0}{x_1}\right) + \frac{r_0^{\lambda}}{\lambda\gamma_1} \quad ,$$

uniformly for

$$z \in \Delta, \qquad |z| \geq r_0(\varepsilon) .$$

Hence we may give to (18.16) the more convenient form

(18.17) $$\log F(z) = B_1 z^\lambda (1+\eta(z)),$$

where $\eta(z) \to 0$, uniformly as $z \to \infty$, $z \in \Delta$.

It is in general impossible to differentiate an asymptotic relation. In the case of (18.17) it is easy to justify differentiation by the following straightforward use of Cauchy's formula for the derivatives

(18.18) $$\frac{d^k}{dz^k} \log F(z) = \frac{k!}{2\pi i} \int_C B_1 \frac{\zeta^\lambda (1+\eta(\zeta))}{(\zeta-z)^{k+1}} d\zeta = B_1 \frac{d^k}{dz^k} z^\lambda + E_1(z) ,$$

where the contour of integration C is given by

$$C : \zeta = z + \kappa|z|e^{i\Theta} \qquad (0 \leq \Theta \leq 2\pi,\ 0 < \kappa < 1) ,$$

and κ is so small that the whole of C lies in the interior of the sector Δ. For the error term $E_1(z)$ we find

(18.19) $$|E_1(z)| \leq B_1 |z|^{\lambda-k} \frac{k!(1+\kappa)^\lambda}{\kappa^k} \max_{\zeta \in C} |\eta(\zeta)| .$$

It is now obvious that (18.17), (18.18) and (18.19) imply

(18.20) $$z \frac{F'(z)}{F(z)} = a(z) = B_1 \lambda z^\lambda (1+\eta(z)),$$

(18.21) $$za'(z) = b(z) = B_1 \lambda^2 z^\lambda (1+\eta(z)),$$

where $\eta(z) \to 0$, uniformly as $z \to \infty$ in the sector Δ.

From (18.21) it follows that the analytic function $a'(z)$ does not vanish identically. Hence, by (18.4), the zeros of $a'(r)$ are isolated and consequently, in view of (18.20), $a(r)$ is strictly increasing and unbounded. We have thus proved assertion I of Theorem 7.

From (18.20) and (18.17)

$$\log F'(z) = \log F(z) + O(\log|z|) = B_1 z^{\lambda}(1+\eta(z))$$

and by a classical result of Laguerre (cf. [1, p. 32]) all the zeros of $F'(z)$ are real and negative. This establishes the fact that: if $F(z)$ is an $\mathcal{L}$-function of genus zero, all its successive derivatives are $\mathcal{L}$-functions of genus zero.

We finally remark that, for $0 < \lambda \leq \frac{1}{2}$, all the zeros of $E_{1/\lambda}(z)$ are real and negative. This result, obtained by Wiman in 1905 [39], was reproved, very simply and convincingly, by Pólya [24]. Now from [39, p. 220]

$$E_{1/\lambda}(r) \sim \lambda \exp(r^{\lambda}) \qquad (r \to +\infty),$$

so that, from (2.53) and (2.54), $E_{1/\lambda}(z)$ $(0 < \lambda \leq \frac{1}{2})$ and all its derivatives are $\mathcal{L}$-functions of genus zero.

19. $\mathcal{L}$-functions of genus zero are admissible in the sense of Hayman. In his remarkable paper [14, pp. 68-69] Hayman introduced a class of functions which he called admissible.

If $F(z)$ is admissible, it must satisfy a number of conditions which need not be stated here. For our purpose, it suffices to note that, if $F(z)$ is an $\mathcal{L}$-function of genus zero, then, by (18.17) and (18.21),

$$\log F(r) \to +\infty, \quad b(r) = ra'(r) \to +\infty \quad (r \to +\infty).$$

Hence Theorem XI [14, p. 88] guarantees that $F(z)$ is admissible and consequently Hayman's fundamental Theorem I [14, p. 69] states that the coefficients a_n, in (2.53) are such that, as

$r \to +\infty$, we have

$$(19.1) \quad a_n r^n = \frac{F(r)}{\{2\pi b(r)\}^{1/2}} \left\{\exp\left[-\frac{(a(r)-n)^2}{2b(r)}\right] + \eta(r,n)\right\} \quad (\eta(r,n) \to 0),$$

uniformly for all integers n.

As indicated by our notational conventions, this means that, given $\varepsilon > 0$, there exists an $r_0(\varepsilon)$, <u>independent of</u> n, such that

$$(19.2) \qquad |\eta(r,n)| < \varepsilon \qquad (r > r_0(\varepsilon)).$$

Hayman draws from (19.1) the following simple consequences [14, p. 75, Theorem II]:

$$(19.3) \qquad \sum_{j \le a(r)} a_j r^j \sim \frac{1}{2} F(r) \qquad (r \to +\infty),$$

$$(19.4) \qquad \sum_{a(r) < j} a_j r^j \sim \frac{1}{2} F(r) \qquad (r \to +\infty).$$

20. <u>The functions</u> $U_m(w)$, $Q_m(w)$, $G_m(w)$ <u>associated with $\mathcal{L}$-functions of genus zero</u>. Since $a(r)$ is nonnegative, continuous, strictly increasing and unbounded, there exists a

uniquely defined, positive, increasing, unbounded sequence $\{R_m\}_m$ such that

$$(20.1) \qquad a(R_m) = m \qquad (m = 1,2,3,\ldots;\ a(0) = 0) \ .$$

The new meaning of the quantities U_m, Q_m, G_m and $b_j(m)$, is the one derived from the expressions (4.6), (4.7), (4.8) and (4.9) with f replaced by an $\mathcal{L}$-function F, and $R = R_m$ defined by (20.1).

As a consequence of these choices, (18.20) and (18.21) yield

$$(20.2) \qquad R_m = \left\{\frac{m}{B_1\lambda}\right\}^{1/\lambda}(1+\eta'_m) \ ,$$

$$(20.3) \qquad b(R_m) = \lambda m(1+\eta''_m) \ ,$$

with

$$(20.4) \qquad |\eta'_m| + |\eta''_m| \to 0 \qquad (m \to +\infty) \ .$$

[We have used the symbols η' in order to suspend, temporarily, our notational convention regarding η .]

From (19.1) and (20.1), we deduce

$$(20.5) \qquad a_m R^m = \frac{F(R)}{\{2\pi b(R)\}^{1/2}}(1+\eta(R,m)) \qquad (R = R_m) \ ,$$

$$(20.6) \qquad a_{m+j}R^{m+j} = \frac{F(R)}{\{2\pi b(R)\}^{1/2}}\left(\exp\left(-\frac{j^2}{2b(R)}\right) + \eta(R,m+j)\right) \ .$$

The relations (4.9) now take the form

$$(20.7) \qquad b_j(m) = \frac{a_{m+j}}{a_m}R^j = \left\{\exp\left(-\frac{j^2}{2b(R)}\right) + \eta(R,m+j)\right\}(1+\eta(R,m))^{-1} \ ,$$

which, in view of (20.3), may be rewritten as

$$(20.8)\qquad b_j(m) = \left(\exp\left(-\frac{j^2}{2\lambda m}\right) + \tilde{\eta}_j(m) + \eta(R,m+j)\right)(1+\eta(R,m))^{-1} \quad ,$$

with

$$(20.9)\qquad |\tilde{\eta}_j(m)| \le \left|\exp\left(\frac{j^2\eta''_m}{2\lambda m(1+\eta''_m)}\right) - 1\right| \quad , \quad j \ge -m \quad .$$

It is clear that, for j fixed, we have, as $m \to +\infty$,

$$\tilde{\eta}_j(m) \to 0, \qquad \eta(R_m, m+j) \to 0, \qquad b_j(m) \to 1 \quad .$$

Hence (4.12) remains valid with the new meaning of the coefficients $b_j(m)$.

From (19.3) and (20.1) we deduce

$$\sum_{j=0}^{m} a_{m-j} R^{m-j} \sim \frac{1}{2} F(R) \qquad (R = R_m,\ m \to +\infty) \quad ,$$

and in view of (20.5), (20.3) and (4.7)

$$Q_m(1) \sim \left(\frac{\pi\lambda m}{2}\right)^{1/2} \qquad (m \to +\infty) \quad ,$$

$$(20.10)\qquad |Q_m(w)| \le (\pi\lambda m)^{1/2} \qquad (|w| \ge 1,\ m > m_0) \quad .$$

Similarly, by (19.4), (20.3), (20.5), and (4.8)

$$G_m(1) \sim \left(\frac{\pi\lambda m}{2}\right)^{1/2} \qquad (m \to +\infty) \quad ,$$

$$(20.11)\qquad |G_m(w)| \le (\pi\lambda m)^{1/2} \qquad (|w| \le 1,\ m > m_0) \quad .$$

Because of its importance in the applications that follow, we draw the attention of the reader to the uniformity in w , of the inequalities (20.10) and (20.11).

21. Estimates for $U_m(w)$. It is convenient to summarize, in our next lemma, some elementary consequences of Cauchy's formula.

Lemma 21.1. *Let* $f(z)$ *be regular, and* $f(z) \neq 0$ *throughout the disk*

$$|z - z_0| \leq 2\eta |z_0| \qquad (0 < \eta < \tfrac{1}{2},\ z_0 \neq 0) \quad . \tag{21.1}$$

Then, if s *is complex and*

$$|s| \leq \eta |z_0| \quad , \tag{21.2}$$

we have

$$\log f(z_0 + s) - \log f(z_0) = s\,\frac{f'(z_0)}{f(z_0)} + \frac{s^2}{2}\left(\frac{f''(z_0)}{f(z_0)} - \left\{\frac{f'(z_0)}{f(z_0)}\right\}^2\right) + E_3(z_0, s), \tag{21.3}$$

where

$$|E_3(z_0, s)| \leq \frac{1}{2(1 - 2\eta)}\,\eta^{-2}\left|\frac{s}{z_0}\right|^3 \max_\theta |a(z_0(1 + 2\eta e^{i\theta}))| \tag{21.4}$$

$$\left(a(z) = z\,\frac{f'(z)}{f(z)}\right) \quad .$$

Proof. Put

$$g(s) = \log\left(\frac{f(z_0 + s)}{f(z_0)}\right) \qquad (g(0) = 0) \quad .$$

By Cauchy's formula

$$g(s) = \frac{1}{2\pi i}\int_C \frac{g(\zeta)}{\zeta - s}\,d\zeta \qquad (|s| \leq \eta |z_0|) \quad ,$$

where the contour C is chosen to be

$$C : \zeta = 2\eta z_0 e^{i\theta} \qquad (0 \leq \theta \leq 2\pi) \quad .$$

By the elements of function theory we conclude that (21.3) holds with

$$E_3(z_0,s) = \frac{s^3}{2\pi i}\int_C \frac{g(\zeta)}{\zeta^3(\zeta-s)}\, d\zeta \quad ,$$

$$(21.5)\qquad |E_3(z_0,s)| \le \frac{1}{4}\left(\frac{|s|}{\eta|z_0|}\right)^3 \max_\theta \left|\log\left(\frac{f(z_0(1+2\eta e^{i\theta}))}{f(z_0)}\right)\right| \quad .$$

To complete the proof of the lemma note that the maximum in the above inequality is

$$\left|\int_{z_0}^{z_0(1+2\eta e^{i\psi})} \frac{a(t)}{t}\, dt\right| \le \frac{2\eta}{1-2\eta} \max_\theta \left|a(z_0(1+2\eta e^{i\theta}))\right| \quad ;$$

the estimate (21.4) clearly follows.

We now apply Lemma 21.1 to $F(z)$ defined by (2.53) and give to the parameters z_0 and s the values

$$(21.6)\qquad z_0 = R_m \ , \qquad s = (w-1)R_m \quad ,$$

with

$$(21.7)\qquad |w-1| \le \eta \qquad (0 < \eta < \tfrac{1}{2}) \quad .$$

In view of (20.1) and (20.3) we see that (21.3) and (21.6) yield

$$(21.8)\qquad F(R_m w) = F(R_m)\exp\left\{(w-1)m + \frac{(w-1)^2}{2} m(\lambda-1+\lambda\eta_m'') + E_3(R_m,(w-1)R_m)\right\} \quad .$$

The restrictions

$$(21.9)\qquad (1-2\eta)R_m \le |z| \le 2R_m \ , \qquad z \in \Delta \quad ,$$

together with (18.20) and (20.2) imply

$$(21.10)\qquad |a(z)| \leq B_1\lambda 2^{\lambda+1} R_m^{\lambda} \leq 2^{\lambda+2} m \qquad (m > m_0) \quad .$$

An estimate for $E_3(R_m,(w-1)R_m)$ follows at once from (21.4) and (21.10):

$$(21.11)\quad |E_3(R_m,(w-1)R_m)| \leq \frac{2^{\lambda+1}\eta^{-2}}{1-2\eta} |w-1|^3 m \qquad (m > m_0) \quad .$$

Combining (4.6), (21.8), (20.5), (20.3) and the approximation

$$(21.12)\quad \log w = -(1-w) - \frac{1}{2}(1-w)^2 + \omega A(1-w)^3 \qquad (|1-w| \leq \tfrac{1}{2}) \quad ,$$

we find

$$(21.13)\quad U_m(w) = (2\pi\lambda m)^{1/2}(1+\eta_m)\exp\left\{\frac{(w-1)^2}{2} m\lambda(1+\eta_m'') + E_4(R_m,(w-1)R_m)\right\},$$

$$(\eta_m \to 0) \quad ,$$

where, by (21.11) and (21.12)

$$(21.14)\qquad |E_4(R_m,(w-1)R_m)| \leq K(\lambda,\eta)m|w-1|^3 \quad .$$

An inspection of (21.13) suggests the introduction of a new complex variable ζ , given exactly as in Lemma 8.1 by the transformation

$$(21.15)\qquad w = 1 + (2/\lambda m)^{1/2}\zeta \qquad (|\zeta| \leq B < +\infty) \quad .$$

[The positive bound B may be chosen arbitrarily.]

The following lemma is now obvious

<u>Lemma 21.2.</u> <u>As</u> $m \to +\infty$,

$$(21.16) \qquad U_m\left(1 + \left(\frac{2}{\lambda m}\right)^{1/2} \zeta\right) (2\pi\lambda m)^{-\frac{1}{2}} \longrightarrow \exp(\zeta^2) \quad ,$$

<u>uniformly on any compact set of the</u> ζ-<u>plane</u>.

From (4.13), (20.11) and (21.16), we see that the condition

$$(21.17) \quad (2\pi\lambda m)^{-\frac{1}{2}} \left| Q_m\left(1 + \left(\frac{2}{\lambda m}\right)^{1/2} \zeta\right)\right| \le 2(\exp(|\zeta|^2) + 1) \quad (m > m_0)$$

is satisfied, uniformly for all ζ such that

$$(21.18) \qquad \left|1 + \left(\frac{2}{\lambda m}\right)^{1/2} \zeta\right| \le 1 \ , \qquad |\zeta| \le B \ .$$

If

$$\left|1 + \left(\frac{2}{\lambda m}\right)^{1/2} \zeta\right| \ge 1 \ ,$$

a direct appeal to (20.10) shows that the inequalities (21.17) are still valid. We may in view of (4.7), (20.5) and (21.15) express (21.17) in terms of the sections $s_m(z)$; this leads immediately to

<u>Lemma 21.3.</u> <u>The functions</u>

$$(21.19) \quad \Omega_m(\zeta) = \left(1 + \left(\frac{2}{\lambda m}\right)^{1/2} \zeta\right)^{-m} \{F(R)\}^{-1} s_m\left(R\left(1 + \left(\frac{2}{\lambda m}\right)^{1/2} \zeta\right)\right)$$

$$(R = R_m),$$

<u>are uniformly bounded on every compact set of the</u> ζ-<u>plane</u>.

22. Determination of $\lim \Omega_m(\zeta)$. In view of Vitali's theorem it will be sufficient to establish the value of the limit for ζ real, positive and

$$1 \le \zeta = x \le B \quad . \tag{22.1}$$

Our first task is to estimate with some precision the coefficients $b_{-j}(m)$ $(0 \le j \le m)$ defined by (4.9) and approximated in (20.7), (20.8) and (20.9).

Introduce the integer

$$L(m) = \left[\frac{m^{1/2}}{|\eta_m''|^{1/3} + (\log m)^{-1}} \right] , \tag{22.2}$$

where η_m'' is defined in (20.3). This choice implies

$$L(m) m^{-\frac{1}{2}} = H_m \to +\infty \qquad (m \to +\infty) , \tag{22.3}$$

as well as

$$L^2(m) |\eta_m''| \le |\eta_m''|^{1/3} m , \qquad H_m < \log m \quad . \tag{22.4}$$

Hence, if

$$|j| \le L(m) , \tag{22.5}$$

we deduce from (20.9) and (22.4)

$$|\tilde{\eta}_j(m)| \le \exp\left(\frac{|\eta_m''|^{1/3}}{\lambda}\right) - 1 = \alpha_m , \quad \alpha_m \to 0 \quad (m \to \infty) \quad . \tag{22.6}$$

The quantities α_m are independent of j, and the inequalities in (22.6) hold uniformly for all j restricted by (22.5).

From (22.6), (20.8) and the fact, expressed in (19.2), that the quantities $\eta(R_m, m+j)$ are uniformly small as $m \to \infty$, we conclude that

$$(22.7)\qquad b_{-j}(m) = \exp\left(-\frac{j^2}{2\lambda m}\right) + \beta_m(-j) \qquad (0 \le j \le L(m))$$

where, uniformly in j ,

$$(22.8)\qquad |\beta_m(-j)| \le \beta'_m \quad , \qquad \beta'_m \to 0 \qquad (m \to +\infty) \quad .$$

For

$$j > L(m) = H_m m^{1/2} \quad ,$$

we deduce from (20.3), (20.7) and (22.3),

$$(22.9)\qquad 0 < b_{-j}(m) \le \left(\exp\left(-\frac{H_m^2}{3\lambda}\right) + \eta(R_m, m-j)\right)(1 + \eta(R_m, m))^{-1} \le \beta''_m \quad ,$$

with

$$(22.10)\qquad \beta''_m \to 0 \qquad (m \to +\infty) \quad .$$

By (22.7) - (22.10) we obtain

$$(22.11)\qquad \left|Q_m(w) - \sum_{j=0}^{L(m)} \exp\left(-\frac{j^2}{2\lambda m}\right) w^{-j}\right| \le (\beta'_m + \beta''_m)\frac{|w|}{|w|-1} \qquad (|w| > 1).$$

The change of variable (21.15) and the restrictions (22.1) show that

$$(22.12)\qquad (2\pi\lambda m)^{-\frac{1}{2}} Q_m\left(1+\left(\frac{2}{\lambda m}\right)^{\frac{1}{2}} x\right) = (2\pi\lambda m)^{-\frac{1}{2}} \sum_{j=0}^{L(m)} \exp\left(-\frac{j^2}{2\lambda m}\right)\left(1+\left(\frac{2}{\lambda m}\right)^{\frac{1}{2}} x\right)^{-j} +$$

$$+\, \omega(\beta'_m + \beta''_m) B\pi^{-\frac{1}{2}} \qquad (1 \le x \le B,\ m > m_0) \quad .$$

Using the approximation

$$\log(1+u) = u + \omega u^2 \qquad (|u| \le \tfrac{1}{2},\ |\omega| \le 1) \quad ,$$

and the second inequality in (22.4), we find that the conditions

$$0 \le x \le B, \qquad m > m_0, \qquad 0 \le j \le L(m) = H_m m^{1/2} \quad ,$$

imply

$$(22.13)\quad \left(1+\left(\frac{2}{\lambda m}\right)^{1/2}x\right)^{-j} = \exp\left(-j\left(\frac{2}{\lambda m}\right)^{1/2}x\right)\left(1+\omega\frac{3x^2 j}{\lambda m}\right)$$

$$= \exp\left(-\left(\frac{2}{\lambda m}\right)^{1/2}jx\right)\left(1+\omega\frac{3B^2\log m}{\lambda m^{1/2}}\right) .$$

Consider now

$$(22.14)\quad \Lambda_m(x) = (2\pi\lambda m)^{-\frac{1}{2}}\sum_{j=0}^{L(m)}\exp\left(-\frac{j^2}{2\lambda m}-\frac{2jx}{(2\lambda m)^{1/2}}\right)$$

$$= (2\pi\lambda m)^{-\frac{1}{2}}\exp(x^2)\sum_{j=0}^{L(m)}\exp\left\{-\left(x+j(2\lambda m)^{-\frac{1}{2}}\right)^2\right\} ,$$

and notice that (22.12), (22.13) and (22.14) imply

$$(22.15)\quad \left|(2\pi\lambda m)^{-\frac{1}{2}}Q_m\left(1+\left(\frac{2}{\lambda m}\right)^{1/2}x\right)-\Lambda_m(x)\right| \le$$

$$\le Km^{-1}\log m\sum_{j=0}^{L(m)}\exp\left(-\frac{j^2}{2\lambda m}\right)+(\beta'_m+\beta''_m)B\pi^{-\frac{1}{2}}\quad (1\le x\le B,\ m>m_0).$$

To complete the evaluation of $\lim_{m\to\infty}\Omega_m(x)$ we compare, in the next section, the last sum in (22.14) with an integral.

23. Comparison with integrals; proof of assertion II of Theorem 7.

Obviously

$$\sum_{j=0}^{L(m)}\exp\left(-\frac{j^2}{2\lambda m}\right)\le 1+\int_0^{+\infty}\exp\left(-\frac{t^2}{2\lambda m}\right)dt = 1+\left(\frac{\pi\lambda m}{2}\right)^{1/2} ,$$

and hence

$$(23.1)\qquad \left\{(2\pi\lambda m)^{-\frac{1}{2}}Q_m\left(1+\left(\frac{2}{\lambda m}\right)^{1/2}x\right)-\Lambda_m(x)\right\}\to 0\qquad (m\to+\infty) ,$$

uniformly on the interval $1 \le x \le B$.

From the elementary inequalities

$$\exp\left\{-\left(x+\frac{j+1}{(2\lambda m)^{1/2}}\right)^2\right\} < \int_j^{j+1} \exp\left\{-\left(x+\frac{t}{(2\lambda m)^{1/2}}\right)^2\right\}dt < \exp\left\{-\left(x+\frac{j}{(2\lambda m)^{1/2}}\right)^2\right\}$$

we conclude that

$$0 < \sum_{j=0}^{L(m)} \exp\left\{-\left(x+\frac{j}{(2\lambda m)^{1/2}}\right)^2\right\} - \int_0^{L(m)+1} \exp\left\{-\left(x+\frac{t}{(2\lambda m)^{1/2}}\right)^2\right\}dt < e^{-x^2} .$$

Returning to (22.14) we see that

$$(23.2) \qquad \left|\Lambda_m(x)-e^{x^2}(2\pi\lambda m)^{-\frac{1}{2}}\int_0^{+\infty} \exp\left\{-\left(x+\frac{t}{(2\lambda m)^{1/2}}\right)^2\right\}dt\right|$$

$$< (2\pi\lambda m)^{-\frac{1}{2}} + e^{x^2}(2\pi\lambda m)^{-\frac{1}{2}} \int_{H_m m^{1/2}}^{+\infty} \exp\left\{-\left(x+\frac{t}{(2\lambda m)^{1/2}}\right)^2\right\}dt .$$

The change of variable

$$\sigma = x + \frac{t}{(2\lambda m)^{1/2}}$$

enables us to replace (23.2) by

$$(23.3) \qquad \left|\Lambda_m(x)-e^{x^2}\pi^{-\frac{1}{2}}\int_x^{+\infty} e^{-\sigma^2}d\sigma\right| < (2\pi\lambda m)^{-\frac{1}{2}} + e^{x^2}\pi^{-\frac{1}{2}}\int_{H_m/(2\lambda)^{1/2}}^{+\infty} e^{-\sigma^2}d\sigma .$$

Since $H_m \to +\infty$ (by (22.3)), the relations (23.1) and (23.3) imply

$$(23.4) \qquad (2\pi\lambda m)^{-\frac{1}{2}} Q_m\left(1+\left(\frac{2}{\lambda m}\right)^{1/2} x\right) \to e^{x^2}\pi^{-\frac{1}{2}}\int_x^{+\infty} e^{-\sigma^2}d\sigma = \frac{e^{x^2}}{2}\operatorname{erfc}(x) \qquad (m \to +\infty) ,$$

uniformly for all $x \in [1,B]$.

Vitali's theorem, Lemma 21.3 and (23.4) yield assertion II of Theorem 7.

24. The Szegö curves for $\mathfrak{L}$-functions of genus zero. Start from (4.6) with w replaced by $te^{i\phi}w$ and let

(24.1) $$0 < t \le 1, \quad 0 < \phi < \pi, \quad R = R_m, \quad m > m_0 ,$$

where R_m is defined in (20.1).

Assume also that

$$|w - 1| < \eta \qquad (0 < \eta < 1) ,$$

and that η is so small that the whole disk

(24.2) $$|z - Rte^{i\phi}| < \eta Rt$$

lies in the sector Δ of (18.10).

Hence (4.6) takes the form

(24.3) $$\log U_m(te^{i\phi}w) = \log F(Rte^{i\phi}w) - \log(a_m R^m) - m \log w - m \log t - im\phi ,$$

which implies

(24.4) $$\Xi_m(t) = \log|U_m(te^{i\phi})| = \log|F(Rte^{i\phi})| - \log(a_m R^m) - m \log t \qquad (R = R_m) .$$

We now prove

Lemma 24.1. Let $F(z)$ satisfy the asymptotic relation (18.17). Then, if $\Xi_m(t)$ is defined by (24.4) and

$$m > m_0 , \quad R = R_m , \quad 0 < \phi < \pi ,$$

it is possible to find $\sigma_m(\phi)$ such that

(24.5) $$\Xi_m(\sigma_m(\phi)) = 0 , \qquad \tilde{\sigma}_0 < \sigma_m(\phi) < \tilde{\sigma}_1 ,$$

where

(24.6) $$\tilde{\sigma}_0 = \exp\left(- \frac{3}{\lambda}\right) , \qquad \tilde{\sigma}_1 = \exp\left(- \frac{1 - \cos\phi\lambda}{2\lambda}\right) .$$

Moreover

$$\sigma_m(\phi) \to \sigma(\phi) \qquad (m \to +\infty) \quad , \tag{24.7}$$

where $x = \sigma(\phi)$ is the unique value of x such that

$$0 < x < 1 \ , \qquad v(x) = 0 \quad , \tag{24.8}$$

with

$$v(x) = x^{\lambda}\cos(\phi\lambda) - 1 - \lambda \log x \quad . \tag{24.9}$$

Remark. If we supplement our definition of $\sigma(\phi)$ by the symmetry relations

$$\sigma(-\phi) = \sigma(\phi) \ , \qquad \sigma(0) = 1 \quad , \tag{24.10}$$

it is clear that $\sigma(\phi)e^{i\phi}$ $(-\pi < \phi < \pi)$ defines a curve which is the analogue, for our new problem, of the Szegö curve introduced in (2.5) and (2.6). The analogy is complete and could be investigated in great detail. We shall be content to study the implications of (24.5); they will be sufficient to lead to assertion III of Theorem 7.

Proof. The function $\Xi_m(t)$ is, for m fixed, a function of the positive variable t which is continuous in the interval

$$\frac{1}{2}\,\tilde{\sigma}_0 \le t \le 1 \quad .$$

Let $\tau \in [\tilde{\sigma}_0/2, 1]$; then, by (18.17)

$$\log|F(R\tau e^{i\phi})| = (1+\eta_m(\tau))\tau^{\lambda}\cos(\phi\lambda)\log F(R) \qquad (R = R_m); \tag{24.11}$$

by (18.17), (18.20) and (20.1)

$$m = (1+\eta_m)\lambda \log F(R) \qquad (R = R_m) \quad , \tag{24.12}$$

and by (20.5) and (20.3)

$$(24.13) \qquad \log(a_m R^m) = \log F(R) - \frac{1}{2}\log(2\pi\lambda m) + \eta_m \qquad (R = R_m) .$$

From (24.4), (24.9), (24.11), (24.12), (24.13) we conclude that

$$(24.14) \qquad \Xi_m(\tau) = (v(\tau) + \eta_m(\tau))\log F(R_m) \qquad (\eta_m(\tau) \to 0)$$

uniformly as $m \to \infty$ and $\tau \in [\tilde{\sigma}_0/2,1]$.

If

$$(24.15) \qquad \exp\left(-\frac{1-\cos(\phi\lambda)}{2\lambda}\right) = \tilde{\sigma}_1 \le \tau \le 1 \quad ,$$

then, by an elementary computation we find

$$(24.16) \qquad v(\tau) \le \frac{-1+\cos(\phi\lambda)}{2} < 0 \qquad (\cos(\phi\lambda) \ge 0) \quad ,$$

and

$$(24.17) \qquad v(\tau) \le \frac{-1-\cos(\phi\lambda)}{2} < 0 \qquad (\cos(\phi\lambda) < 0) \quad .$$

Similarly

$$(24.18) \qquad v(\tilde{\sigma}_0) \ge 1 \quad .$$

Using (24.18) and (24.16) or (24.17) in (24.14) we find

$$(24.19) \quad \Xi_m(\tilde{\sigma}_0) > 0 \ , \quad \Xi_m(\tau) < 0 \qquad (\tilde{\sigma}_1 \le \tau \le 1, \ \ m > m_0) \quad .$$

Hence because of the continuity of $\Xi_m(t)$ it is possible, given $\phi (0 < \phi < \pi)$, to find $m_0(\phi)$ and $\sigma_m(\phi)$ such that, for $m > m_0(\phi)$, the conditions (24.5) are satisfied.

Since we cannot guarantee the monotone character of $\Xi_m(t)$, there may be several values of $t \in [\tilde{\sigma}_0,\tilde{\sigma}_1]$ for which $\Xi_m(t)$ vanishes. In order to make our choice of σ_m unique, we impose the additional restriction

$$\Xi_m(t) < 0 \qquad (\sigma_m < t \le 1) \quad ;$$

in view of (24.19) this is certainly possible.

From (24.14) we deduce

$$v(\sigma_m) + \eta_m(\sigma_m) = 0 \qquad (m > m_0) \quad ,$$

$$\lim_{m\to\infty} v(\sigma_m) = 0 \quad .$$

To complete the proof of the lemma it suffices to note that $v(x)$ is strictly decreasing and hence $\sigma(\phi)$ in (24.7) is uniquely determined by the conditions

$$v(\sigma(\phi)) = 0 \quad , \quad 0 < \sigma(\phi) < 1 \quad .$$

The latter relation implies assertion III(i) of Theorem 7.

25. Estimates for $U_m(\sigma_m e^{i\phi} w)$. Apply Lemma 21.1 with

$$z_0 + s = R_m\sigma_m e^{i\phi} w \quad , \quad z_0 = R_m\sigma_m e^{i\phi} \quad , \quad |w-1| \le \eta \qquad (0 < \eta < \tfrac{1}{2}) \tag{25.1}$$

where $\sigma_m = \sigma_m(\phi)$. We shall be content with the approximation

$$\log F(R_m\sigma_m e^{i\phi} w) = \log F(R_m\sigma_m e^{i\phi}) + (w-1)a(R_m\sigma_m e^{i\phi}) + E_5 \quad , \tag{25.2}$$

where, as in the derivation of (21.14), we have

$$|E_5| \le K|w-1|^2 m \qquad (0 < K) \quad , \tag{25.3}$$

and $K = K(\lambda,\eta)$ is independent of m.

Combining (4.6) (with w replaced by $\sigma_m e^{i\phi} w$) and (25.2) we find

$$\log U_m(\sigma_m e^{i\phi} w) = \log F(R\sigma_m e^{i\phi}) - \log(a_m R^m) - m\log\sigma_m - im\phi \tag{25.4}$$

$$- m\log w + (w-1)a(R\sigma_m e^{i\phi}) + E_5 \qquad (R = R_m) \quad .$$

Our definition of $\Xi_m(t)$ (in (24.4)) and (24.5) show that the sum of the first three terms in the right-hand side of (25.4) is

$$(25.5) \qquad i \arg F(R\sigma_m e^{i\phi}) \quad .$$

The information regarding (25.5) to be derived from (18.17) is not precise enough to be useful. For our purpose it suffices to define χ_m such that

$$(25.6) \qquad -\pi \le \chi_m < \pi \ , \qquad \chi_m \equiv \arg F(R\sigma_m e^{i\phi}) - m\phi \qquad (\text{mod } 2\pi)$$

and to observe that the introduction of χ_m and the obvious consequence of (18.20) and (20.2):

$$a(R\sigma_m e^{i\phi}) = m\sigma_m^\lambda e^{i\phi\lambda}(1+\eta_m) \quad ,$$

enable us to deduce from (25.4)

$$U_m(\sigma_m e^{i\phi} w) e^{-i\chi_m} = w^{-m} \exp((w-1)m\sigma_m^\lambda e^{i\phi\lambda}(1+\eta_m) + E_5) \quad ,$$

$$|E_5| \le K|w-1|^2 m \qquad (|w-1| \le \tfrac{1}{2}) \quad .$$

The change of variable

$$w-1 = \frac{\zeta}{m(1-\xi^\lambda)} \qquad (\xi = \sigma(\phi)e^{i\phi})$$

shows that, as $m \to +\infty$,

$$(25.7) \qquad \exp(-i\chi_m)U_m\left(\sigma_m e^{i\phi}\left(1+\frac{\zeta}{m(1-\xi^\lambda)}\right)\right) \to e^{-\zeta}$$

uniformly on every compact set of the ζ-plane. Assertion III (ii) of Theorem 7 follows from the above relation together with (4.12) and (4.13) (which are still valid).

26. Proof of assertion IV of Theorem 7. By (25.6) $\chi_m \in [-\pi,\pi)$; hence, from any unbounded sequence of positive, increasing integers it is possible to extract an infinite subsequence such that

$$\chi_m \to \chi \qquad (m \to +\infty,\ m \in M) \quad . \tag{26.1}$$

Then (25.7) yields

$$U_m\left(\sigma_m(\phi)e^{i\phi}\left(1 + \frac{\zeta}{m(1-\xi^\lambda)}\right)\right) \longrightarrow e^{i\chi}\, e^{-\zeta} \quad . \tag{26.2}$$

We have noticed that (4.12) and (4.13) are valid. Since $\lim_{m\to\infty} \sigma_m(\phi) = \sigma(\phi) < 1$, we see that (26.2) implies, as $m \to +\infty$, $m \in M$,

$$Q_m\left(\sigma_m(\phi)e^{i\phi}\left(1 + \frac{\zeta}{m(1-\xi^\lambda)}\right)\right) \longrightarrow e^{i\chi}\, e^{-\zeta} - \frac{\sigma(\phi)}{e^{-i\phi}-\sigma(\phi)} \tag{26.3}$$

uniformly on every compact subset of the ζ-plane.

Returning to (4.7) we find from (20.3) and (20.5)

$$Q_m\left(\sigma_m(\phi)e^{i\phi}\left(1 + \frac{\zeta}{m(1-\xi^\lambda)}\right)\right) \sim \tag{26.4}$$

$$\frac{(2\pi\lambda m)^{1/2}}{F(R_m)}\, s_m\left(R_m\sigma_m(\phi)e^{i\phi}\left(1 + \frac{\zeta}{m(1-\xi^\lambda)}\right)\right)\{\sigma_m(\phi)\}^{-m} e^{-i\phi m} \exp\left(\frac{-\zeta}{1-\xi^\lambda}\right)$$

$$(m \to +\infty,\ m \in M) \ .$$

This completes the proof of assertion IV of Theorem 7.

References

1. E. Borel, Leçons sur les fonctions entières, 2nd ed., Gauthier-Villars, Paris, 1921.
2. J. D. Buckholtz, "A characterization of the exponential series," Amer. Math. Monthly 73(1966), 121-123.
3. F. Carlson, "Sur les fonctions entières," C. R. Acad. Sci. Paris 179(1924), 1583-1585.
4. F. Carlson, "Sur les fonctions entières," Ark. Mat. Astron. Fys. 35A, No. 14(1948), 18 pp.
5. M. Cartwright, Integral Functions, Cambridge University Press, Cambridge, 1956.
6. J. Dieudonné, "Sur les zéros des polynomes--sections de e^x," Bull. Sci. Math. 70(1935), 333-351.
7. A. Edrei, "Solution of the deficiency problem for functions of small lower order," Proc. London Math. Soc. 26(1973), 435-445.
8. A. Edrei, "The Padé tables of entire functions," J. Approx. Theory, 28(1980), 54-82.
9. A. Edrei, "On a conjecture of Saff and Varga," to appear.
10. A. Edrei and W. H. J. Fuchs, "The deficiencies of meromorphic functions of order less than one," Duke Math. J. 27(1960), 223-250.
11. H. E. Fettis, J. C. Caslin, and K. R. Cramer, "Complex zeros of the error function and of the complementary error function," Math. Comp. 27(1973), 401-404.

12. T. Ganelius, "The zeros of the partial sums of power series," Duke Math. J. 30(1963), 533-540.

13. G. H. Hardy and J. E. Littlewood, "Notes on the theory of series (XI): On Tauberian theorems," Proc. London Math. Soc. 30(1930), 23-37.

14. W. K. Hayman, "A generalisation of Stirling's formula," Journal für r. und angew. Math., 196(1956), 67-95.

15. K. E. Iverson, "The zeros of the partial sums of e^z," Math. Tables Aids Comp. 7(1953), 163-168.

16. R. Jentzsch, "Untersuchungen zur Theorie der Folgen analytischer Funktionen," Acta Math. 41(1917), 219-251.

17. B. Ja. Levin, Distribution of zeros of entire functions, Translations of Mathematical Monographs, Vol. 5, Amer. Math. Soc., Providence, 1964.

18. E. Lindelöf, "Mémoire sur la théorie des fonctions entieres de genre fini," Acta Soc. Scient. Fennicae, 31(1902), 1-79.

19. J. Malmquist, "Étude d'une fonction entière," Acta Math. 29(1905), 203-215.

20. R. Nevanlinna, Le Théorème de Picard-Borel et la Théorie des Fonctions Méromorphes, Gauthier-Villars, Paris, 1929.

21. D. J. Newman and T. J. Rivlin, "The zeros of the partial sums of the exponential function," J. Approx. Theory 5(1972), 405-412.

22. D. J. Newman and T. J. Rivlin, "Correction: the zeros of the partial sums of the exponential function," J. Approx. Theory 16(1976), 299-300.

23. A. Pflüger, "Die Wertverteilung und das Verhalten von Betrag und Argument einer speziellen Klasse analytischer Funktionen," Comment. Math. Helv. 11(1938), 180-214.

24. G. Pólya, "Bemerkung über die Mittag-Lefflerschen Funktionen $E_\alpha(z)$," Tôhoku Math. J., 19(1921), 241-248.

25. G. Pólya and G. Szegö, Problems and Theorems in Analysis II, Springer-Verlag, Berlin, 1976.

26. P. C. Rosenbloom, Sequences of Polynomials, Especially Sections of Power Series, Ph.D. Thesis, Stanford University, 1943.

27. P. C. Rosenbloom, "Distribution of zeros of polynomials," in Lectures on Functions of a Complex Variable (W. Kaplan, ed.), University of Michigan Press, Ann Arbor, 1955, pp. 265-285.

28. E. B. Saff and R. S. Varga, "On the zeros and poles of Padé approximants to e^z," Numer. Math. 25(1975), 1-14.

29. E. B. Saff and R. S. Varga, "Zero-free parabolic regions for sequences of polynomials," SIAM J. Math. Anal. 7(1976), 344-357.

30. E. B. Saff and R. S. Varga, "Geometric overconvergence of rational functions in unbounded domains," Pacific J. Math. 62(1976), 523-549.

31. E. B. Saff and R. S. Varga, "Some open problems concerning polynomials and rational functions," Padé and Rational Approximations: Theory and Applications (E. B. Saff and R. S. Varga, eds.), pp. 483-488, Academic Press, Inc., New York, 1977.

32. E. B. Saff and R. S. Varga, "On the zeros and poles of Padé approximants to e^z. III," Numer. Math. 30(1978), 241-266.

33. G. Szegö, "Über die Nullstellen von Polynomen die in einem Kreise gleichmässig konvergieren," Sitzungsber. Berl. Math. Ges. 21(1922), 59-64.

34. G. Szegö, "Über eine Eigenschaft der Exponentialreihe," Sitzungsber. Berl. Math. Ges. 23(1924), 50-64.

35. E. C. Titchmarsh, The Theory of the Riemann Zeta-Function, Oxford, at the Clarendon Press, 1951.

36. E. C. Titchmarsh, The Theory of Functions, 2nd ed., Oxford Univ. Press, London and New York, 1939.

37. G. Valiron,"Sur les fonctions entières d'ordre fini et d'ordre nul, et en particulier les fonctions à correspondance régulière,"Ann. Fac. Sci. Univ. Toulouse, 5(1913), 117-257.

38. R. S. Varga, Properties of a Special Set of Entire Functions and Their Respective Partial Sums, Ph.D. Thesis, Harvard University, 1954.

39. A. Wiman, "Über die Nullstellen der Funktionen $E_\alpha(x)$," Acta Math., 29(1905), 217-234.

Index of Ad-Hoc Definitions and Notations

Basic decomposition:
$Q_m(w) = U_m(w) - G_m(w)$, 42

$b_j(m)$, 42, 93

$E_{1/\lambda}(z)$, 3

$G_m(w)$, 42, 92
    estimates for, 49, 51

$J_m(w)$, 53
    differential equation for, 54
    estimates for, 60

L, 47, 99

$\mathfrak{L}$-function, 20

$Q_m(w)$, 42, 92
    estimates for, 68

R_m, 7, 93

S, 8, 9

$S_\phi(\tau)$, 5

$U_m(w)$, 42, 92
    estimates for, 63-69
    for $\mathfrak{L}$-functions, 95-98, 106-107

$X(\cdot)$, 62

λ, 2

η_j, 44

ω, 43

General Index

admissible function (of Hayman) 91-92
asymptotic relations (for Mittag-Leffler functions) 44-45

Buckholtz, J.D. 4, 109

Carlson, F. 2, 5, 25, 109
central index 8, 42
complementary error function 10, 11, 12, 18, 57
    zeros of 11, 18, 58-59, 72

Dieudonné, J. 4, 109

Edrei, A. 3, 7, 25, 109
error function 58
exponential function 2, 3, 4

Ganelius, T. 12, 24, 110

Hardy, G.H. 21, 110
Hayman, W.K. 91, 110

Iverson, K.E. 4, 110

Jentzsch, R. 1, 110

$\mathcal{L}$-function (of genus zero) 3, 19, 24, 28
    derivatives of 91
    partial sums of 21-23
    properties of 87-91
    zeros of 88-91
Levin, B. Ja. 25, 110
Lindelöf, E. 20, 110
Littlewood, J.E. 21, 110

Malmquist, J. 24, 110
maximum term 8, 42
Mittag-Leffler function 3
    asymptotic properties of 45
    partial sums of 10-19

Nevanlinna, R. 21, 58, 110
Newman, D.J. 4, 110
notational conventions 43, 44
numerical results 26, 29-40

open problems 23-25

Padé table 25
parabolic region 4, 18, 23
partial sums 1
 of $\mathcal{L}$-function 21-22
 of Mittag-Leffler functions 10-19, 26
Pfluger, A. 25, 110
Pólya, G. 24, 91, 111

Rivlin, T.J. 4, 110
Rosenbloom, P.C. 2, 5, 25, 111

Saff, E.B. 4, 25, 111
Stirling's formula 46
Szegö, G. 1, 2, 24, 112
Szegö's curve, 8, 9, 12, 14, 16, 17, 18, 27, 28, 62
 circular portion 9, 14, 77
 for $\mathcal{L}$-functions 103 - 104

tauberian theorem (of Valiron) 88

Valiron, G. 8, 20, 21, 88, 112
Varga, R.S. 5, 25, 111, 112

Width Conjecture 4, 5, 19, 23, 24, 26
 Modified 5, 6, 18, 24, 26
Wiman, A. 8, 21, 91, 112

zeros
 Hurwitz 27, 28
 spurious 2, 28
zero free regions for partial sums
 for exp(z) 4, 5
 for Mittag-Leffler functions 16-19, 26-28